The Stable Manifold Theorem for Semilinear Stochastic Evolution Equations and Stochastic Partial Differential Equations

MEMOIRS
of the
American Mathematical Society

Number 917

The Stable Manifold Theorem for Semilinear Stochastic Evolution Equations and Stochastic Partial Differential Equations

Salah-Eldin A. Mohammed
Tusheng Zhang
Huaizhong Zhao

November 2008 • Volume 196 • Number 917 (fourth of 5 numbers) • ISSN 0065-9266

American Mathematical Society
Providence, Rhode Island

2000 *Mathematics Subject Classification.*
Primary 60H10, 60H20; Secondary 60H25.

Library of Congress Cataloging-in-Publication Data

Mohammed, Salah-Eldin A., 1946–
The stable manifold theorem for semilinear stochastic evolution equations and stochastic partial differential equations / Salah-Eldin A. Mohammed, Tusheng Zhang, Huaizhong, Zhao.
p. cm. — (Memoirs of the American Mathematical Society, ISSN 0065-9266 ; no. 917)
"Volume 196, number 917 (fourth of 5 numbers)."
Includes bibliographical references.
ISBN 978-0-8218-4250-8 (alk. paper)
1. Stochastic partial differential equations. 2. Stochastic integral equations. 3. Manifolds (Mathematics) 4. Evolution equations. I. Zhang, Tusheng, 1963– II. Zhao, Huaizhong, 1964– III. Title.

QA274.25.M64 2008
519.2—dc22 2008030290

Memoirs of the American Mathematical Society

This journal is devoted entirely to research in pure and applied mathematics.

Subscription information. The 2008 subscription begins with volume 191 and consists of six mailings, each containing one or more numbers. Subscription prices for 2008 are US$675 list, US$540 institutional member. A late charge of 10% of the subscription price will be imposed on orders received from nonmembers after January 1 of the subscription year. Subscribers outside the United States and India must pay a postage surcharge of US$38; subscribers in India must pay a postage surcharge of US$43. Expedited delivery to destinations in North America US$53; elsewhere US$130. Each number may be ordered separately; *please specify number* when ordering an individual number. For prices and titles of recently released numbers, see the New Publications sections of the *Notices of the American Mathematical Society.*

Back number information. For back issues see the *AMS Catalog of Publications.*

Subscriptions and orders should be addressed to the American Mathematical Society, P. O. Box 845904, Boston, MA 02284-5904, USA. *All orders must be accompanied by payment.* Other correspondence should be addressed to 201 Charles Street, Providence, RI 02904-2294, USA.

Memoirs of the American Mathematical Society (ISSN 0065-9266) is published bimonthly (each volume consisting usually of more than one number) by the American Mathematical Society at 201 Charles Street, Providence, RI 02904-2294, USA. Periodicals postage paid at Providence, RI. Postmaster: Send address changes to Memoirs, American Mathematical Society, 201 Charles Street, Providence, RI 02904-2294, USA.

This publication is indexed in *Science Citation Index®*, *SciSearch®*, *Research Alert®*, *CompuMath Citation Index®*, *Current Contents®/Physical, Chemical & Earth Sciences.*
Printed in the United States of America.

∞ The paper used in this book is acid-free and falls within the guidelines established to ensure permanence and durability.
Visit the AMS home page at `http://www.ams.org/`

10 9 8 7 6 5 4 3 2 1 13 12 11 10 09 08

Contents

Introduction 1

Part 1. The stochastic semiflow

§1.1 Basic concepts 6

§1.2 Flows and cocycles of semilinear see's 7

(a) Linear see's 8

(b) Semilinear see's 27

§1.3 Semilinear spde's: Lipschitz nonlinearity 35

§1.4 Semilinear spde's: Non-Lipschitz nonlinearity 44

(a) Stochastic reaction diffusion equations 44

(b) Burgers equation with additive noise 58

Part 2. Existence of stable and unstable manifolds

§2.1 Hyperbolicity of a stationary trajectory 69

§2.2 The nonlinear ergodic theorem 73

§2.3 Proof of the local stable manifold theorem 77

§2.4 The local stable manifold theorem for see's and spde's 95

(a) See's: Additive noise 95

(b) Semilinear see's: Linear noise 98

(c) Semilinear parabolic spde's: Lipschitz nonlinearity 100

(d) Stochastic reaction diffusion equations: Dissipative nonlinearity 100

(e) Stochastic Burgers equation: Additive noise 101

Acknowledgments 102

Bibliography 103

ABSTRACT. The main objective of this paper is to characterize the pathwise local structure of solutions of semilinear stochastic evolution equations (see's) and stochastic partial differential equations (spde's) near stationary solutions. Such characterization is realized through the long-term behavior of the solution field near stationary points. The analysis falls in two parts 1, 2. In Part 1, we prove general existence and compactness theorems for C^k-cocycles of semilinear see's and spde's. Our results cover a large class of semilinear see's as well as certain semilinear spde's with Lipschitz and non-Lipschitz terms such as stochastic reaction diffusion equations and the stochastic Burgers equation with additive infinite-dimensional noise. In Part 2, stationary solutions are viewed as cocycle-invariant random points in the infinite-dimensional state space. The pathwise local structure of solutions of semilinear see's and spde's near stationary solutions is described in terms of the almost sure longtime behavior of trajectories of the equation in relation to the stationary solution. More specifically, we establish *local stable manifold theorems* for semilinear see's and spde's (Theorems 2.4.1-2.4.4). These results give smooth stable and unstable manifolds in the neighborhood of a hyperbolic stationary solution of the underlying stochastic equation. The stable and unstable manifolds are stationary, live in a stationary tubular neighborhood of the stationary solution and are asymptotically invariant under the stochastic semiflow of the see/spde. Furthermore, the local stable and unstable manifolds intersect transversally at the stationary point, and the unstable manifolds have fixed finite dimension. The proof uses infinite-dimensional multiplicative ergodic theory techniques, interpolation and perfection arguments (Theorem 2.2.1).

Received by the editor 7/8/05.

1991 *Mathematics Subject Classification.* Primary 60H10, 60H20; Secondary 60H25.

Key words and phrases. Stochastic semiflow, C^k cocycle, stochastic evolution equation (see), stochastic partial differential equation (spde), multiplicative ergodic theorem, stationary solution, hyperbolicity, local stable (unstable) manifolds..

The research of the first author is supported in part by NSF Grants DMS-9703852, DMS-9975462 and DMS-0203368.
The research of the second author is supported in part by EPSRC Grant GR/R91144.
The research of the third author is supported in part by EPSRC Grants GR/R69518 and GR/R93582.

Introduction

The construction of local stable and unstable manifolds near hyperbolic equilibria is a fundamental problem in deterministic and stochastic dynamical systems. The significance of these invariant manifolds consists in a characterization of the local behavior of the dynamical system in terms of longtime asymptotics of its trajectories near a stationary point. In recent years, it has been established that local stable/unstable manifolds exist for finite-dimensional stochastic ordinary differential equations (sode's) ([M-S.2]) and stochastic systems with finite memory (viz. stochastic functional differential equations (sfde's))([M-S.1]). On the other hand, existence of such manifolds for nonlinear stochastic evolution equations (see's) and stochastic partial differential equations (spde's) has been an open problem since the early nineties ([F-S], [B-F], [B-F.1]).

In [F-S], the existence of a random evolution operator and its Lyapunov spectrum is established for a linear stochastic heat equation on a bounded Euclidean domain, driven by finite-dimensional white noise. For linear see's with finite-dimensional white noise, a stochastic semi-flow (i.e. random evolution operator) was obtained in [B-F]. Subsequent work on the dynamics of nonlinear spde's has focused mainly on the question of existence of *continuous* semiflows and the existence and uniqueness of invariant measures and/or stationary solutions. Recent results on the existence of global invariant, stable/unstable manifolds (through a fixed point) for semilinear see's are given in ([D-L-S.1],[D-L-S.2]). The results in ([D-L-S.1], [D-L-S.2]) assume that the see is driven by multiplicative one-dimensional Brownian motion, with the nonlinear term having a global Lipschitz constant that is sufficiently small relative to the spectral gaps of the second-order linear operator. The latter spectral gap condition in ([D-L-S.1], [D-L-S.2]) is dictated by the use of the contraction mapping theorem.

The main objective of this article is to establish the existence of local stable and unstable manifolds near stationary solutions of semilinear stochastic evolution equations (see's) and stochastic partial differential equations (spde's). Our approach consists in the following two major undertakings:

- A construction of a sufficiently Fréchet differentiable, locally compact cocycle for mild/weak trajectories of the see or the spde. Part 1 of this paper is devoted to detailing the construction of the cocycle.
- The application of classical nonlinear ergodic theory techniques developed by Oseledec [O] and Ruelle [Ru.2] in order to study the local structure of the above cocycle in a neighborhood of a hyperbolic stationary point. This structure characterizes-via stable/unstable manifolds-the asymptotic stability of the cocycle near the stationary point.

The problem of existence of semiflows for see's and spde's is a nontrivial one, mainly due to the well-established fact that finite-dimensional methods for constructing (even continuous) stochastic flows break down in the infinite-dimensional setting of spde's and see's. In particular, Kolmogorov's continuity theorem fails for random fields parametrized by infinite-dimensional Hilbert spaces ([Mo.1], pp. 144-149, [Sk], [Mo.2]). (Cf. also [F.1], [F.2], [D-Z.1], pp. 246-248). In view of the failure of Kolmogorov's theorem, issues of perfection of the infinite-dimensional semiflow and its ergodic properties are of prime importance because it is necessary to extend Ruelle's discrete-time multiplicative ergodic theory to a continuous time setting. On the other hand, there is a significant body of literature on the existence of perfect finite-dimensional stochastic flows and cocycles generated by stochastic ordinary differential equations in Euclidean space or finite-dimensional manifolds. For more details on the existence and regularity of finite-dimensional stochastic flows, the reader may refer to [C], [I-W], [Ku], [A-S], [M-S.4], [M-S.3], [A], [M-S.2], and the references therein. Needless to say that, in these works, Kolmogorov's continuity theorem plays a central role in the construction of the underlying finite-dimensional stochastic flows.

In Part 1 of this article, we show the existence of smooth locally compact perfect cocycles for mild solutions of semilinear see's and spde's. This is achieved in the see case by a combination of a chaos-type expansion and suitable lifting and variational techniques, and for spde's by using stochastic variational representations and methods from deterministic pde's. More specifically, for see's in Hilbert space, our construction employs a "chaos-type" representation in the Hilbert-Schmidt operators, using the linear terms of the see (Theorems 1.2.1-1.2.4). This technique bypasses the need for Kolmogorov's continuity theorem and appears to be new. A variational technique is then employed in order to handle the nonlinear terms (Theorem 1.2.6). Applications to specific classes of spde's are given. In particular, we obtain smooth stochastic semiflows for semilinear spde's driven by cylindrical Brownian motion (Theorem 1.3.5). In these applications, it turns out that in addition to smoothness of the nonlinear terms, one requires some level of dissipativity or Lipschitz continuity in order to guarantee the existence of smooth globally defined semiflows. Specific examples of spde's include semilinear parabolic spde's with Lipschitz nonlinearities (Theorem 1.3.5), stochastic reaction diffusion equations (Theorems 1.4.1, 1.4.2) and stochastic Burgers equations with additive infinite-dimensional noise (Theorem 1.4.3).

As indicated above, the existence of a smooth semiflow is a necessary tool for constructing local stable and unstable manifolds for see's and spde's near a hyperbolic stationary random point, ala work of Oseledec-Ruelle ([O], [Ru.1], [Ru.2]). The construction of these stable/unstable manifolds is the main objective of the analysis in Part 2 of this article. Stationary points correspond to stationary solutions of the see/spde, and are random points invariant under the perfect cocycle constructed in Part 1. Hyperbolicity of a stationary point is defined via the Lyapunov spectrum of the linearization of the cocycle. Using Kingman's subadditive ergodic theorem and Ruelle's discrete nonlinear multiplicative ergodic theory techniques, appropriate estimates are developed in order to control the excursions of the nonlinear cocycle between discrete time points. Thus, perfect versions of the local stable and unstable manifolds are constructed within a stationary neighborhood of the hyperbolic equilibrium. Furthermore, it is shown in this part that the local stable/unstable manifolds are transversal and asymptotically invariant under the nonlinear cocycle. The unstable manifolds are finite-dimensional with fixed (non-random) dimension. These results are referred to collectively as *local stable manifold theorems*. Local stable manifold theorems are established for various classes of semilinear see's and spde's (Theorems 2.4.1-2.4.4). In particular, our results cover semilinear see's, stochastic parabolic equations, stochastic reaction-diffusion equations, and Burgers equation with additive infinite-dimensional noise in $L^2([0,1])$.

From the viewpoint of spde's, the results in this article go well beyond standard issues of existence, uniqueness of mild/weak solutions, existence and ergodicity of the invariant measure. These fundamental issues have occupied the spde community for a long time and are well-studied (cf. for example [Wa], [D-Z.1], [D-Z.2], [Si.1], [Si.2], [C-K-S]). Since the main objective of the present article is to characterize "generic" a.s. local behavior of the stochastic semiflow near equilibria, emphasis is placed largely on issues of hyperbolicity of the stationary solutions rather than existence and uniqueness/ergodicity of the invariant measure. From a dynamical systems point of view, it is needless to say that ergodicity of the stationary solution is a non-generic property. In general, as in the deterministic case, finite and infinite-dimensional stochastic dynamical systems admit more than one stationary point. The stochastic dynamics in a regime with multiple equilibria is not well-understood.

Finally, it should be noted that the case of *nonlinear* multiplicative noise is largely open: It is not known to us if see's driven by nonlinear multidimensional white noise admit perfect (smooth, or even continuous) cocycles. The issue of existence of stationary solutions for spde's driven by nonlinear white noise and their relation to backward sde's is being investigated in [Z-Z].

Part 1

The Stochastic Semiflow

1.1. Basic concepts

This part of our article is devoted to the contruction of Fréchet smooth stochastic semiflows for mild and weak solutions of semilinear see's and spde's.

In Theorem 1.2.6, it is shown that mild solutions of semilinear see's in a Hilbert space H generate smooth perfect locally compacting cocycles. The construction of the cocycle for semilinear see's is based on the following new strategy, which bypasses the need for Kolmogorov's continuity theorem:

- We "lift" the linear terms of the see to the Hilbert-Schmidt operators $L_2(H)$.
- We represent the mild solution of the linear see as a "chaos-type" series expansion living in the Hilbert space $L_2(H)$ of Hilbert-Schmidt operators on H (Theorems 1.2.1-1.2.3).
- Using a variational technique, the mild solution of the full semilinear see is represented in terms of the linear cocycle constructed above (Theorems 1.2.1-1.2.4). This part of the strategy requires the non-linear part of the see to satisfy a Lipschitz condition (Theorem 1.2.6).

Similar variational techniques are used to construct smooth cocycles for weak solutions of specific classes of spde's. In particular, we obtain smooth stochastic semiflows for semilinear spde's driven by cylindrical Brownian motion. In these applications, it turns out that in addition to smoothness of the non-linear terms, one requires some level of dissipativity or Lipschitz continuity, e.g. the stochastic heat equation (Theorem 1.3.5), the stochastic reaction diffusion equation (Theorems 1.4.1, 1.4.2) and stochastic Burgers equation with additive infinite-dimensional noise (Theorem 1.4.3).

We begin by formulating the ideas of a *stochastic semiflow* and a *cocycle* which are central to the analysis in this work.

Let $(\Omega, \mathcal{F}, P)$ be a probability space. Denote by $\bar{\mathcal{F}}$ the P-completion of $\mathcal{F}$, and let $(\Omega, \bar{\mathcal{F}}, (\mathcal{F}_t)_{t\geq 0}, P)$ be a complete filtered probability space satisfying the usual conditions ([Pr]). Denote $\Delta := \{(s,t) \in \mathbf{R}^2 : 0 \leq s \leq t\}$, and $\mathbf{R}^+ := [0, \infty)$. For a topological space E, let $\mathcal{B}(E)$ denote its Borel σ-algebra.

Let k be a positive integer and $0 < \epsilon \leq 1$. If E and N are real Banach spaces, we will denote by $L^{(k)}(E, N)$ the Banach space of all k-multilinear maps $A : E^k \to N$ with the uniform norm $\|A\| := \sup\{|A(v_1, v_2, \cdots, v_k)| : v_i \in E, |v_i| \leq 1, i = 1, \cdots, k\}$. Suppose $U \subseteq E$ is an open set. A map $f : U \to N$ is said to be *of class* $C^{k,\epsilon}$ if it is C^k and if $D^{(k)}f : U \to L^{(k)}(E, N)$ is ϵ-Hölder continuous on bounded sets in U. A $C^{k,\epsilon}$ map $f : U \to N$ is said to be *of class* $C_b^{k,\epsilon}$ if all its derivatives $D^{(j)}f, 1 \leq j \leq k$, are globally bounded on U, and $D^{(k)}f$ is ϵ-Hölder continuous on U. A mapping $\tilde{f} : [0,T] \times U \to N$ is *of class* $C^{k,\epsilon}$ *in the second variable uniformly with respect to the first* if for each $t \in [0,T]$, $\tilde{f}(t, \cdot)$ is $C^{k,\epsilon}$ on U, for every bounded set $U_0 \subseteq U$ the spatial partial derivatives $D^{(j)}\tilde{f}(t,x), j = 1, \cdots, k$, are uniformly bounded in $(t,x) \in [0,T] \times U_0$ and the corresponding ϵ-Hölder constant of $D^{(k)}\tilde{f}(t, \cdot)|U_0$ is uniformly bounded in $t \in [0,T]$.

The following definitions are crucial to the developments in this article.

DEFINITION 1.1.1. Let E be a Banach space, k a non-negative integer and $\epsilon \in (0,1]$. A *stochastic* $C^{k,\epsilon}$ *semiflow* on E is a random field $V : \Delta \times E \times \Omega \to E$ satisfying the following properties:

(i) V is $(\mathcal{B}(\Delta) \otimes \mathcal{B}(E) \otimes \mathcal{F}, \mathcal{B}(E))$-measurable.

(ii) For each $\omega \in \Omega$, the map $\Delta \times E \ni (s,t,x) \mapsto V(s,t,x,\omega) \in E$ is continuous.

(iii) For fixed $(s,t,\omega) \in \Delta \times \Omega$, the map $E \ni x \mapsto X(s,t,x,\omega) \in E$ is $C^{k,\epsilon}$.
(iv) If $0 \le r \le s \le t$, $\omega \in \Omega$ and $x \in E$, then

$$V(r,t,x,\omega) = V(s,t,V(r,s,x,\omega),\omega).$$

(v) For all $(s,x,\omega) \in \mathbf{R}^+ \times E \times \Omega$, one has $V(s,s,x,\omega) = x$.

DEFINITION 1.1.2. Let $\theta : \mathbf{R} \times \Omega \to \Omega$ be a P-preserving $(\mathcal{B}(\mathbf{R}) \otimes \mathcal{F}, \mathcal{F})$-measurable group on the probability space $(\Omega, \mathcal{F}, P)$, E a Banach space, k a non-negative integer and $\epsilon \in (0,1]$. A $C^{k,\epsilon}$ *perfect cocycle* (U,θ) on E is a $(\mathcal{B}(\mathbf{R}^+) \otimes \mathcal{B}(E) \otimes \mathcal{F}, \mathcal{B}(E))$-measurable random field $U : \mathbf{R}^+ \times E \times \Omega \to E$ with the following properties:
(i) For each $\omega \in \Omega$, the map $\mathbf{R}^+ \times E \ni (t,x) \mapsto U(t,x,\omega) \in E$ is continuous; and for fixed $(t,\omega) \in \mathbf{R}^+ \times \Omega$, the map $E \ni x \mapsto U(t,x,\omega) \in E$ is $C^{k,\epsilon}$.
(ii) $U(t+s,\cdot,\omega) = U(t,\cdot,\theta(s,\omega)) \circ U(s,\cdot,\omega)$ for all $s,t \in \mathbf{R}^+$ and all $\omega \in \Omega$.
(iii) $U(0,x,\omega) = x$ for all $x \in E, \omega \in \Omega$.

Note that a cocycle (U,θ) corresponds to a one-parameter semigroup on $E \times \Omega$, viz.

$$\begin{aligned} \mathbf{R}^+ \times E \times \Omega &\to E \times \Omega \\ (t,(x,\omega)) &\mapsto (U(t,x,\omega),\theta(t,\omega)) \end{aligned}$$

Fig.1 illustrates the cocycle property. The vertical solid lines represent random copies of E sampled according to the probability measure P.

The main objective of this part of our article is to show that under sufficient regularity conditions on the coefficients, a large class of semilinear see's and spde's admits a $C^{k,\epsilon}$ semiflow $V : \Delta \times H \times \Omega \to H$ for a suitably chosen state space H with the following property: For every $x \in H$, $V(t_0,\cdot,x,\cdot)$ coincides a.s. for all $t \ge t_0$ with the mild/weak solution of the see/spde with initial function x at $t = t_0$. In the autonomous case, we show further that the semiflow V generates a cocycle (U,θ) on H, in the sense of Definition 1.1.2 above. The cocycle and its Fréchet derivative are compact in all cases.

1.2. Flows and cocycles of semilinear see's

In this section, we will establish the existence and regularity of semiflows generated by mild solutions of semilinear see's. We will begin with the linear case. In fact, the linear cocycle will be used to represent the mild solution of the semilinear see via a variational formula which transforms the semilinear see to a random integral equation (Theorem 1.2.5). The latter equation plays a key role in establishing the regularity of the stochastic flow of the semilinear see (Theorem 1.2.6).

One should note at this point the fact that Kolmogorov's continuity theorem fails for random fields parametrized by infinite-dimensional spaces. As a simple example, consider the random field $I : L^2([0,1],\mathbf{R}) \to L^2(\Omega,\mathbf{R})$ defined by the Wiener integral

$$I(x) := \int_0^1 x(t)\, dW(t), \quad x \in L^2([0,1],\mathbf{R}),$$

where W is one-dimensional Brownian motion. The above random field has no continuous (or even linear!) measurable selection $L^2([0,1],\mathbf{R}) \times \Omega \to \mathbf{R}$ ([Mo.1], pp. 144-148; [Mo.2]).

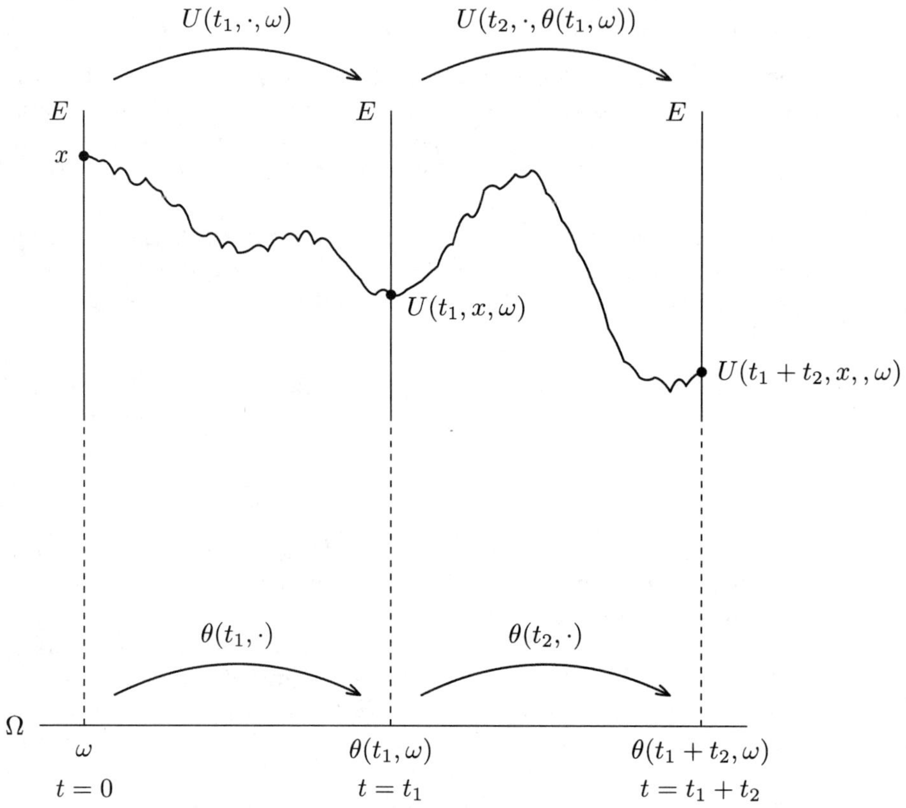

FIGURE 1. The Cocycle Property.

(a) *Linear see's*

We will first prove the existence of semiflows associated with mild solutions of linear stochastic evolution equations of the form:

$$\left.\begin{aligned} du(t,x,\cdot) &= -\,Au(t,x,\cdot)dt + Bu(t,x,\cdot)\,dW(t), \quad t>0 \\ u(0,x,\omega) &= x \in H. \end{aligned}\right\} \tag{1.2.1}$$

In the above equation $A : D(A) \subset H \to H$ is a closed linear operator on a separable real Hilbert space H. Assume that A has a complete orthonormal system of eigenvectors $\{e_n : n \geq 1\}$ with corresponding positive eigenvalues $\{\mu_n, n \geq 1\}$; i.e., $Ae_n = \mu_n e_n$, $n \geq 1$. Suppose $-A$ generates a strongly continuous semigroup of bounded linear operators $T_t : H \to H$, $t \geq 0$. Let E be a separable Hilbert space and $W(t), t \geq 0$, be an E-valued Brownian motion with a separable covariance Hilbert space K, and defined on the canonical complete filtered Wiener space $(\Omega, \bar{\mathcal{F}}, (\mathcal{F}_t)_{t\geq 0}, P)$, satisfying the usual conditions. Here $K \subset E$ is a Hilbert-Schmidt embedding. Indeed, Ω is the space of all continuous paths $\omega : \mathbf{R} \to E$ such that $\omega(0) = 0$, furnished with the compact open topology; $\mathcal{F}$ is its Borel σ-field; P is Wiener measure on Ω; $\bar{\mathcal{F}}$ is the P-completion of $\mathcal{F}$; and $\mathcal{F}_t$ is the P-completion of the sub-σ-field of $\mathcal{F}$ generated by all evaluations $\Omega \ni \omega \mapsto \omega(u) \in E, u \leq t$. The Brownian motion is given by

$$W(t,\omega) := \omega(t), \quad \omega \in \Omega,\, t \in \mathbf{R},$$

and may be represented by

$$W(t) = \sum_{k=1}^{\infty} W^k(t) f_k, \quad t \in \mathbf{R},$$

where $\{f_k : k \geq 1\}$ is a complete orthonormal basis of K, and $W^k, k \geq 1$, are standard independent one-dimensional Wiener processes ([D-Z.1], Chapter 4). Note that, in general, the above series converges in E but not in K.

Let $L(K,H)$ be the Banach space of all bounded linear maps $T : K \to H$ given the uniform operator norm

$$\|T\|_{L(K,H)} := \sup_{\substack{x \in K \\ |x| \leq 1}} |T(x)|.$$

As usual, we let $L(H)$ be the Banach space of all bounded linear operators $H \to H$ given the uniform operator norm $\|\cdot\|_{L(H)}$.

Denote by $L_2(K,H) \subset L(K,H)$ the Hilbert space of all Hilbert-Schmidt operators $S : K \to H$, given the norm

$$\|S\|_2 := \left[\sum_{k=1}^{\infty} |S(f_k)|^2\right]^{1/2},$$

where $|\cdot|$ is the norm on H. In particular, $L_2(H) \subset L(H)$ stands for the Hilbert space of all Hilbert-Schmidt operators $S : H \to H$. It is easy to see that if $S \in L_2(H)$ and $T \in L(H)$, then $\|S\|_{L(H)} \leq \|S\|_2$, $T \circ S$ (and $S \circ T$) $\in L_2(H)$ and $\|T \circ S\|_{L_2(H)} \leq \|T\|_{L(H)} \|S\|_{L_2(H)}$.

In the see (1.2.1), we assume that $B : H \to L_2(K,H)$ is a bounded linear operator. The stochastic integral in (1.2.1) is defined in the following sense ([D-Z.1], Chapter 4):

Let $F : [0,a] \times \Omega \to L_2(K,H)$ be $\mathcal{B}([0,a] \otimes \mathcal{F}, \mathcal{B}(L_2(K,H)))$-measurable, $(\mathcal{F}_t)_{t\geq 0}$-adapted and such that $\int_0^a E\|F(t)\|^2_{L_2(K,H)}\, dt < \infty$. Define

$$\int_0^a F(t)\, dW(t) := \sum_{k=1}^{\infty} \int_0^a F(t)(f_k)\, dW^k(t)$$

where the H-valued stochastic integrals on the right hand side are with respect to the one-dimensional Wiener processes W^k, $k \geq 1$. Note that the above series converges in $L^2(\Omega, H)$ because

$$\sum_{k=1}^{\infty} E\left|\int_0^a F(t)(f_k)\, dW^k(t)\right|^2 = \int_0^a E\|F(t)\|^2_{L_2(K,H)}\, dt < \infty.$$

Throughout the rest of the article, we will denote by $\theta : \mathbf{R} \times \Omega \to \Omega$ the standard P-preserving ergodic Wiener shift on Ω:

$$\theta(t, \omega)(s) := \omega(t + s) - \omega(t), \quad t, s \in \mathbf{R}.$$

Hence (W, θ) is a *helix*:

$$W(t_1 + t_2, \omega) - W(t_1, \omega) = W(t_2, \theta(t_1, \omega)), \quad t_1, t_2 \in \mathbf{R}, \, \omega \in \Omega.$$

A *mild solution* of (1.2.1) is a family of $(\mathcal{B}(\mathbf{R}^+) \otimes \mathcal{F}, \mathcal{B}(H))$-measurable, $(\mathcal{F}_t)_{t \geq 0}$-adapted processes $u(\cdot, x, \cdot) : \mathbf{R}^+ \times \Omega \to H, \; x \in H$, satisfying the following stochastic integral equation:

$$u(t, x, \cdot) = T_t x + \int_0^t T_{t-s} B u(s, x, \cdot) \, dW(s), \quad t \geq 0. \tag{1.2.2}$$

The next lemma describes a canonical lifting of the strongly continuous semigroup $T_t : H \to H, t \geq 0$, to a strongly continuous semigroup of bounded linear operators $\tilde{T}_t : L_2(K, H) \to L_2(K, H), t \geq 0$.

LEMMA 1.2.1. *Define the family of maps* $\tilde{T}_t : L_2(K, H) \to L_2(K, H), t \geq 0$, *by*

$$\tilde{T}_t(C) := T_t \circ C, \quad C \in L_2(K, H), \, t \geq 0.$$

Then the following is true:
(i) $\tilde{T}_t$, $t \geq 0$, *is a strongly continuous semigroup of bounded linear operators on* $L_2(K, H)$; *and* $\|\tilde{T}_t\|_{L(L_2(K,H))} = \|T_t\|_{L(H)}$ *for all* $t \geq 0$.
(ii) *If* $\tilde{A} : \mathcal{D}(\tilde{A}) \subset L_2(K, H) \to L_2(K, H)$ *is the infinitesimal generator of* $\tilde{T}_t$, $t \geq 0$, *then*

$$\mathcal{D}(\tilde{A}) = \{C : C \in L_2(K, H), \, C(K) \subseteq \mathcal{D}(A), A \circ C \in L_2(K, H)\}$$

and

$$\tilde{A}(C) = A \circ C$$

for all $C \in \mathcal{D}(\tilde{A})$.
(iii) $\tilde{T}_t, t \geq 0$, *is a contraction semigroup if* $T_t, t \geq 0$, *is.*

PROOF. Observe that each $\tilde{T}_t : L_2(K, H) \to L_2(K, H), t \geq 0$, is a bounded linear map of $L_2(K, H)$ into itself. Indeed, it is easy to see that

$$\|\tilde{T}_t(C)\|_{L_2(K,H)} \leq \|T_t\|_{L(H)} \|C\|_{L_2(K,H)}, \quad C \in L_2(K, H), \, t \geq 0; \tag{1.2.3}$$

and hence $\|\tilde{T}_t\|_{L(L_2(K,H))} \leq \|T_t\|_{L(H)}$ for all $t \geq 0$. This implies assertion (iii). The reverse inequality

$$\|T_t\|_{L(H)} \leq \|\tilde{T}_t\|_{L(L_2(K,H))}, \quad t \geq 0,$$

is not hard to check. Hence the last assertion in (i) holds.

We next verify the semi-group property of $\tilde{T}_t, t \geq 0$. Let $t_1, t_2 \geq 0, C \in L_2(K, H)$. Then

$$(\tilde{T}_{t_2} \circ \tilde{T}_{t_1})(C) = T_{t_2} \circ (T_{t_1} \circ C) = T_{t_1 + t_2} \circ C = \tilde{T}_{t_1 + t_2}(C).$$

Note also that $\tilde{T}_0 = I_{L(L_2(K,H))}$, the identity map $L_2(K,H) \to L_2(K,H)$. Therefore, $\tilde{T}_t$, $t \geq 0$, is a semigroup on $L_2(K,H)$. To prove the strong continuity of $\tilde{T}_t$, $t \geq 0$, we will show that

$$\lim_{t \to 0+} \tilde{T}_t(C) = C \tag{1.2.4}$$

for each $C \in L_2(K,H)$. To prove the above relation, let $C \in L_2(K,H)$ and recall that $\{f_k : k \geq 1\}$ is a complete orthonormal basis of K. From the strong continuity of $T_t, t \geq 0$, it follows that

$$\lim_{t \to 0+} |T_t(C(f_k)) - C(f_k)|_H^2 = 0 \tag{1.2.5}$$

for each integer $k \geq 1$. Furthermore,

$$|T_t(C(f_k)) - C(f_k)|_H^2 \leq 2[\sup_{0 \leq t \leq a} \|T_t\|_{L(H)}^2 + 1]|C(f_k)|_H^2, \quad k \geq 1. \tag{1.2.6}$$

Since C is Hilbert-Schmidt, (1.2.6) implies that the series

$$\sum_{k=1}^{\infty} |T_t(C(f_k)) - C(f_k)|_H^2$$

converges uniformly with respect to t. Therefore, from (1.2.5), (1.2.6) and dominated convergence, it follows that

$$\begin{aligned} \lim_{t \to 0+} \|\tilde{T}_t(C) - C\|_{L_2(K,H)}^2 &= \lim_{t \to 0+} \sum_{k=1}^{\infty} |T_t(C(f_k)) - C(f_k)|_H^2 \\ &= \sum_{k=1}^{\infty} \lim_{t \to 0+} |T_t(C(f_k)) - C(f_k)|_H^2 = 0. \end{aligned} \tag{1.2.7}$$

Therefore, (1.2.4) holds and $\tilde{T}_t, t \geq 0$, is strongly continuous.

We next prove assertion (ii) of the lemma. Let $\tilde{A} : \mathcal{D}(\tilde{A}) \subset L_2(K,H) \to L_2(K,H)$ be the infinitesimal generator of $\tilde{T}_t, t \geq 0$. We begin with a proof of the inclusion

$$\{C : C \in L_2(K,H),\, C(K) \subseteq \mathcal{D}(A), A \circ C \in L_2(K,H)\} \subseteq \mathcal{D}(\tilde{A}). \tag{1.2.8}$$

Let $C \in L_2(K,H)$ be such that $C(K) \subseteq \mathcal{D}(A)$ and $A \circ C \in L_2(K,H)$. We will show that

$$\lim_{t \to 0+} \frac{\tilde{T}_t(C) - C}{t} = A \circ C \tag{1.2.9}$$

in $L_2(K,H)$. To prove (1.2.9), note first that

$$\begin{aligned} \sup_{0 \leq t \leq a} \frac{1}{t}|T_t(C(f_k)) - C(f_k)|_H &= \sup_{0 \leq t \leq a} \frac{1}{t}\left|\int_0^t T_s(A(C(f_k)))\,ds\right|_H \\ &\leq \sup_{0 \leq t \leq a} \|T_s\|_{L(H)}|A(C(f_k))|_H \end{aligned} \tag{1.2.10}$$

because $C(f_k) \in \mathcal{D}(A)$ for every $k \geq 1$. Since

$$\|A \circ C\|_{L_2(K,H)} = \sum_{k=1}^{\infty} |A(C(f_k))|_H^2 < \infty, \tag{1.2.11}$$

it follows from (1.2.10), (1.2.11) and dominated convergence that

$$
\begin{aligned}
\limsup_{t\to 0+}&\left\|\frac{\tilde{T}_t(C)-C}{t}-A\circ C\right\|^2_{L_2(K,H)} \\
&=\limsup_{t\to 0+}\sum_{k=1}^{\infty}\left|\frac{T_t(C(f_k))-C(f_k)}{t}-A(C(f_k))\right|^2_H \\
&\leq \sum_{k=1}^{\infty}\limsup_{t\to 0+}\left|\frac{T_t(C(f_k))-C(f_k)}{t}-A(C(f_k))\right|^2_H \qquad (1.2.12)\\
&=0.
\end{aligned}
$$

This proves (1.2.9). In particular, $C\in\mathcal{D}(\tilde{A})$ and $\tilde{A}(C)=A\circ C$.

It remains to prove the inclusion

$$\mathcal{D}(\tilde{A})\subseteq\{C: C\in L_2(K,H),\, C(K)\subseteq\mathcal{D}(A), A\circ C\in L_2(K,H)\}. \tag{1.2.13}$$

Suppose $C\in\mathcal{D}(\tilde{A})$. We will show that $C(K)\subseteq\mathcal{D}(A)$, $A\circ C\in L_2(K,H)$ and $\tilde{A}(C)=A\circ C$. Since

$$
\begin{aligned}
\lim_{t\to 0+}&\left\|\frac{\tilde{T}_t(C)-C}{t}-\tilde{A}(C)\right\|^2_{L_2(K,H)} \\
&=\lim_{t\to 0+}\sum_{k=1}^{\infty}\left|\frac{T_t(C(f_k))-C(f_k)}{t}-\tilde{A}(C)(f_k)\right|^2_H=0, \qquad (1.2.14)
\end{aligned}
$$

we have that

$$\lim_{t\to 0+}\left|\frac{T_t(C(f_k))-C(f_k)}{t}-\tilde{A}(C)(f_k)\right|^2_H=0 \tag{1.2.15}$$

for every $k\geq 1$. Therefore, $C(f_k)\in\mathcal{D}(A)$ and $\tilde{A}(C)(f_k)=A(C(f_k))$ for each $k\geq 1$. Now pick any $f\in K$ and write

$$f^n:=\sum_{k=1}^{n}<f,f_k>f_k,\quad n\geq 1.$$

Then $C(f^n)=\sum_{k=1}^{n}<f,f_k>C(f_k)\in\mathcal{D}(A)$, $n\geq 1$, and $C(f)=\lim_{n\to\infty}C(f^n)$ in H. Now since $\tilde{A}(C)\in L_2(K,H)\subseteq L(K,H)$, it follows that

$$
\begin{aligned}
\tilde{A}(C)(f)&=\lim_{n\to\infty}\tilde{A}(C)(f^n)\\
&=\lim_{n\to\infty}\sum_{k=1}^{n}<f,f_k>\tilde{A}(C)(f_k)\\
&=\lim_{n\to\infty}\sum_{k=1}^{n}<f,f_k>A(C(f_k))\\
&=\lim_{n\to\infty}A(C(f^n)).
\end{aligned}
$$

Since A is a closed operator, the above relation implies that $C(f)\in\mathcal{D}(A)$ and $A(C(f))=\tilde{A}(C)(f)$. As $\tilde{A}(C)\in L_2(K,H)$, and $f\in K$ is arbitrary, it follows that

$C(K) \subseteq \mathcal{D}(A)$, $A \circ C \in L_2(K, H)$ and $\tilde{A}(C) = A \circ C$. This proves (1.2.13) and completes the proof of the lemma. $\square$

Our main results in this section give regular versions $u : \mathbf{R}^+ \times H \times \Omega \to H$ of mild solutions of (1.2.1) such that $u(t, \cdot, \omega) \in L(H)$ for all $(t, \omega) \in \mathbf{R}^+ \times \Omega$ (Theorems 1.2.1-1.2.3). These regular versions are shown to be $L(H)$-valued cocycles with respect to the Brownian shift θ (Theorem 1.2.4). In order to formulate these regularity results, we will require the following lemma:

LEMMA 1.2.2. *Let $B : H \to L_2(K, H)$ be continuous linear, and $v : \mathbf{R}^+ \times \Omega \to L_2(H)$ be a $(\mathcal{B}(\mathbf{R}^+) \otimes \mathcal{F}, \mathcal{B}(H))$-measurable, $(\mathcal{F}_t)_{t \geq 0}$-adapted process such that $\int_0^a E\|v(t)\|^2_{L_2(H)}\, dt < \infty$ for each $a > 0$. Then the random field $\int_0^t \tilde{T}_{t-s}(\{[B \circ v(s)](x)\})\, dW(s)$, $x \in H$, $t \geq 0$, admits a jointly measurable version which will be denoted by $\int_0^t T_{t-s} Bv(s)\, dW(s)$ (by abuse of notation) and has the following properties:*

(i) $\left[\int_0^t T_{t-s} Bv(s)\, dW(s)\right](x) = \int_0^t \tilde{T}_{t-s}(\{[B \circ v(s)](x)\})\, dW(s)$ *for all $x \in H$, $t \geq 0$, a.s.*

(ii) *For a.a. $\omega \in \Omega$ and each $t \geq 0$, the map*

$$H \ni x \mapsto \left[\left(\int_0^t T_{t-s} Bv(s)\, dW(s)\right)(\omega)\right](x) \in H$$

is Hilbert-Schmidt.

PROOF. To prove the lemma, we will define $\int_0^t T_{t-s} Bv(s)\, dW(s)$ as an Itô stochastic integral with values in the Hilbert space $L_2(H)$ in the sense of [D-Z.1], Chapter 4). To do this, we will introduce the following notation.

For any $V \in L_2(H)$ and $B \in L(H, L_2(K, H))$, define the linear map $B \star V : K \to L_2(H)$ by

$$(B \star V)(f)(x) := B(V(x))(f), \quad f \in K, x \in H. \tag{1.2.16}$$

Then $B \star V \in L_2(K, L_2(H))$ because of the following computation

$$\begin{aligned}
\|B \star V\|^2_{L_2(K, L_2(H))} &= \sum_{k=1}^{\infty} \|(B \star V)(f_k)\|^2_{L_2(H)} \\
&= \sum_{k=1}^{\infty} \sum_{n=1}^{\infty} |(B \star V)(f_k)(e_n)|^2_H \\
&= \sum_{k=1}^{\infty} \sum_{n=1}^{\infty} |B(V(e_n))(f_k)|^2 \\
&= \sum_{n=1}^{\infty} \sum_{k=1}^{\infty} |B(V(e_n))(f_k)|^2 \\
&= \sum_{n=1}^{\infty} \|B(V(e_n))\|^2_{L_2(K, H)}
\end{aligned}$$

$$\leq \|B\|^2_{L(H,L_2(K,H))}\|V\|^2_{L_2(H)} < \infty.$$

Now let $v : \mathbf{R}^+ \times \Omega \to L_2(H)$ be as in the lemma. Denote by $\tilde{\tilde{T}}_t : L_2(K, L_2(H)) \to L_2(K, L_2(H))$ the induced lifting of $T_t : H \to H,\ t \geq 0$, via Lemma 1.2.1; i.e.

$$\tilde{\tilde{T}}_t(C)(f) := T_t \circ C(f), \quad C \in L_2(K, L_2(H)),\ f \in K.$$

Fix $t \in [0, a]$. Then the process $[0, t] \ni s \mapsto \tilde{\tilde{T}}_{t-s}(B \star v(s)) \in L_2(K, L_2(H))$ is $(\mathcal{F}_s)_{0 \leq s \leq t}$-adapted and square-integrable, viz.

$$E \int_0^t \|\tilde{\tilde{T}}_{t-s}(B \star v(s))\|^2_{L_2(K,L_2(H))}\, ds$$
$$\leq \|B\|^2_{L(H,L_2(K,H))} \sup_{0 \leq u \leq t} \|T_u\|^2_{L(H)} \int_0^t E\|v(s)\|^2_{L_2(H)}\, ds < \infty.$$

In view of this, the $L_2(H)$-valued Itô stochastic integral $\int_0^t \tilde{\tilde{T}}_{t-s}(B \star v(s))\, dW(s)$ is well-defined ([D-Z.1], Chapter 4). For simplicity of notation, we will denote this stochastic integral by

$$\int_0^t T_{t-s} Bv(s)\, dW(s) := \int_0^t \tilde{\tilde{T}}_{t-s}(B \star v(s))\, dW(s). \tag{1.2.17}$$

This gives the required version of the random field $\int_0^t \tilde{T}_{t-s}(\{[B \circ v(s)](x)\})\, dW(s)$, $x \in H,\ t \geq 0$, because

$$\begin{aligned}
\left[\int_0^t T_{t-s} Bv(s)\, dW(s)\right](x) &:= \left[\int_0^t \tilde{\tilde{T}}_{t-s}(B \star v(s))\, dW(s)\right](x) \\
&= \sum_{k=1}^{\infty} \int_0^t \left[\tilde{\tilde{T}}_{t-s}(B \star v(s))(f_k)\right](x)\, dW^k(s) \\
&= \sum_{k=1}^{\infty} \int_0^t T_{t-s}\{B \star v(s))(f_k)(x)\}\, dW^k(s) \\
&= \sum_{k=1}^{\infty} \int_0^t T_{t-s}\{B(v(s)(x))(f_k)\}\, dW^k(s) \\
&= \sum_{k=1}^{\infty} \int_0^t \tilde{T}_{t-s}\{[B \circ v(s)](x)\}(f_k)\, dW^k(s) \\
&= \int_0^t \tilde{T}_{t-s}\{[B \circ v(s)](x)\}\, dW(s)
\end{aligned}$$

for all $x \in H$ and $t \geq 0$ a.s. In the above computation, we have used the fact that for fixed $x \in H$, the Itô stochastic integral commutes with the continuous linear evaluation map $L_2(H) \ni T \mapsto T(x) \in H$. □

THEOREM 1.2.1. *Assume that for some* $\alpha \in (0,1)$, $A^{-\alpha}$ *is trace-class, i.e.,* $\sum_{n=1}^{\infty} \mu_n^{-\alpha} < \infty$. *Then the mild solution of the linear stochastic evolution equation* (1.2.1) *has a* $(\mathcal{B}(\mathbf{R}^+) \otimes \mathcal{B}(H) \otimes \mathcal{F}, \mathcal{B}(H))$*-measurable version* $u : \mathbf{R}^+ \times H \times \Omega \to H$ *with the following properties:*

(i) *For each* $x \in H$, *the process* $u(\cdot, x, \cdot) : \mathbf{R}^+ \times \Omega \to H$ *is* $(\mathcal{B}(\mathbf{R}^+) \otimes \mathcal{F}, \mathcal{B}(H))$*-measurable,* $(\mathcal{F}_t)_{t\geq 0}$*-adapted and satisfies the stochastic integral equation* (1.2.2).

(ii) *For almost all* $\omega \in \Omega$, *the map* $[0,\infty) \times H \ni (t,x) \to u(t,x,\omega) \in H$ *is jointly continuous. Furthermore, for any fixed* $a \in \mathbf{R}^+$,

$$E \sup_{0\leq t\leq a} \|u(t,\cdot,\cdot)\|_{L(H)}^{2p} < \infty,$$

whenever $p \in (1, \alpha^{-1}]$.

(iii) *For almost all* $\omega \in \Omega$ *and each* $t > 0$, $u(t,\cdot,\omega) : H \to H$ *is a Hilbert-Schmidt operator with the following representation:*

$$u(t,\cdot,\cdot) = T_t + \sum_{n=1}^{\infty} \int_0^t T_{t-s_1} B \int_0^{s_1} T_{s_1-s_2} B \cdots$$

$$\cdots \int_0^{s_{n-1}} T_{s_{n-1}-s_n} B T_{s_n}\, dW(s_n) \cdots dW(s_2)\, dW(s_1). \tag{1.2.18}$$

In the above equation, the iterated Itô stochastic integrals are interpreted in the sense of Lemma 1.2.2, *and the convergence of the series holds in the Hilbert space* $L_2(H)$ *of Hilbert-Schmidt operators on* H.

(iv) *For almost all* $\omega \in \Omega$, *the path* $[0,\infty) \ni t \mapsto u(t,\cdot,\omega) - T_t \in L_2(H)$ *is continuous. In particular, the path* $(0,\infty) \ni t \mapsto u(t,\cdot,\omega) \in L_2(H)$ *is continuous for a.a.* $\omega \in \Omega$. *Furthermore, the process* $u : (0,\infty) \times \Omega \to L_2(H)$ *is* $(\mathcal{F}_t)_{t\geq 0}$*-adapted and* $(\mathcal{B}((0,\infty)) \otimes \mathcal{F}, \mathcal{B}(L_2(H)))$*-measurable.*

PROOF. Under the hypotheses on A, it is well known that the see (1.2.1) has a unique $(\mathcal{F}_t)_{t\geq 0}$-adapted mild solution u satisfying the integral equation (1.2.2) in H. Let $S \subset H$ be a bounded set in H. Using (1.2.2) and a simple application of the Itô isometry together with Gronwall's lemma implies that

$$\sup_{\substack{0\leq t\leq a\\ x\in S}} E[|u(t,x,\cdot)|^2] < \infty$$

for each $a \in (0,\infty)$. Fix $x \in S$ and $t > 0$. Applying (1.2.2) recursively, we obtain by induction

$$\begin{aligned}
u(t,x,\cdot)&\\
=&T_t x + \sum_{k=1}^{n} \Big[\int_0^t T_{t-s_1} B \int_0^{s_1} T_{s_1-s_2} B \cdots \int_0^{s_{k-1}} T_{s_{k-1}-s_k} B T_{s_k}\, dW(s_k) \\
&\qquad \cdots dW(s_2)\, dW(s_1) \Big] x \\
&+ \int_0^t T_{t-s_1} B \int_0^{s_1} T_{s_1-s_2} B \cdots \int_0^{s_n} T_{s_n - s_{n+1}} B u(s_{n+1}, x, \cdot)\, dW(s_{n+1}) \\
&\qquad \cdots dW(s_2)\, dW(s_1)
\end{aligned}$$

for $t>0$. Set $C_t := \sup_{0\le s\le t} \|T_sB\|^2_{L(H,L_2(K,H))}$ for each $t>0$. Therefore,

$$\begin{aligned}
&E\Big[\Big|\int_0^t T_{t-s_1}B\int_0^{s_1} T_{s_1-s_2}B\cdots\int_0^{s_n} T_{s_n-s_{n+1}}Bu(s_{n+1},x,\cdot)dW(s_{n+1})\\
&\qquad\qquad\cdots dW(s_2)\,dW(s_1)\Big|^2\Big]\\
&=\int_0^t ds_1 E\Big[\Big\|T_{t-s_1}B\int_0^{s_1} T_{s_1-s_2}B\cdots\int_0^{s_n} T_{s_n-s_{n+1}}Bu(s_{n+1},x,\cdot)dW(s_{n+1})\\
&\qquad\qquad\cdots dW(s_2)\Big\|^2_{L_2(K,H)}\Big]\\
&\le C_t\int_0^t ds_1 E\Big[\Big|\int_0^{s_1} T_{s_1-s_2}B\cdots\int_0^{s_n} T_{s_n-s_{n+1}}Bu(s_{n+1},x,\cdot)dW(s_{n+1})\\
&\qquad\qquad\cdots dW(s_2)\Big|^2\Big]\\
&\le\cdots\cdots\\
&\le C_t^n\int_0^t ds_1\int_0^{s_1} ds_2\cdots\int_0^{s_n} E[|u(s_{n+1},x,\cdot)|^2]ds_{n+1}\le C_t^n M\frac{t^n}{n!}\to 0,
\end{aligned}$$

where M is a positive constant independent of $x\in S$. This gives the following series representation of $u(t,x,\cdot)$:

$$\begin{aligned}
u(t,x,\cdot) &= T_tx+\sum_{n=1}^{\infty}\Big[\int_0^t T_{t-s_1}B\int_0^{s_1} T_{s_1-s_2}B\cdots\int_0^{s_{n-1}}\\
&\qquad T_{s_{n-1}-s_n}BT_{s_n}\,dW(s_n)\cdots dW(s_2)\,dW(s_1)\Big]x
\end{aligned}\tag{1.2.19}$$

for each $t>0$ and $x\in H$. The above series of iterated Itô stochastic integrals converges in $L^2(\Omega,H)$ uniformly in compacta in t and for x in bounded sets in H.

Using the fact that A^{-1} is trace class, we will show further that the series expansion (1.2.18) actually holds in the Hilbert space $L^2(\Omega,L_2(H))$. To see this, first observe that T_t and all the terms in the series on the right hand side of (1.2.18) are Hilbert-Schmidt for any fixed $t>0$. We use the comparison test to conclude that the series on the right hand side of (1.2.18) converges (absolutely) in $L^2(\Omega,L_2(H))$. Fix $a>0$. Then by successive applications of the Itô isometry (in $L_2(H)$), one gets

$$\begin{aligned}
&E\Big\|\int_0^a T_{a-s_1}B\int_0^{s_1} T_{s_1-s_2}B\cdots\int_0^{s_n} T_{s_n-s_{n+1}}BT_{s_{n+1}}dW(s_{n+1})\\
&\qquad\qquad\cdots dW(s_2)\,dW(s_1)\Big\|^2_{L_2(H)}\\
&=\int_0^a ds_1 E\Big[\Big\|T_{a-s_1}B\int_0^{s_1} T_{s_1-s_2}B\cdots\int_0^{s_n}\\
&\qquad T_{s_n-s_{n+1}}BT_{s_{n+1}}dW(s_{n+1})\cdots dW(s_2)\Big\|^2_{L_2(K,L_2(H))}\Big]
\end{aligned}$$

$$\leq C_a \int_0^a ds_1 E\left[\left\|\int_0^{s_1} T_{s_1-s_2}B\cdots\int_0^{s_n} T_{s_n-s_{n+1}}BT_{s_{n+1}}dW(s_{n+1}) \cdots dW(s_2)\right\|^2_{L_2(H)}\right]$$

$$\leq \cdots$$

$$\leq C_a^n \int_0^a ds_1 \int_0^{s_1} ds_2 \cdots \int_0^{s_n} E[\|BT_{s_{n+1}}\|^2_{L_2(K,L_2(H))}]ds_{n+1}$$

$$\leq C_a^n \frac{a^n}{n!}\int_0^a \|T_s\|^2_{L_2(H)}\, ds = C_a^n \frac{a^n}{n!}\int_0^a \sum_{k=1}^\infty e^{-2\mu_k s}\, ds$$

(1.2.20)
$$\leq C_a^n \frac{a^n}{n!}\sum_{k=1}^\infty \frac{1}{2\mu_k},$$

for each integer $n \geq 1$. This implies that the expansion (1.2.18) converges in $L^2(\Omega, L_2(H))$ for each $t > 0$. Hence assertion (iii) of the theorem holds.

We next prove assertion (iv). Consider the series in (1.2.18) and let $\Phi^n(t) \in L_2(H)$ be its general term, viz.

$$\Phi^n(t) := \int_0^t T_{t-s_1}B\int_0^{s_1} T_{s_1-s_2}B\cdots\int_0^{s_{n-1}} T_{s_{n-1}-s_n}BT_{s_n}\, dW(s_n) \cdots dW(s_2)\, dW(s_1),$$

for $t \geq 0, n \geq 1$. Note the relations

(1.2.21)
$$\left.\begin{aligned} \Phi^n(t) &= \int_0^t T_{t-s_1}B\Phi^{n-1}(s_1)\, dW(s_1), \quad n \geq 2,\\ \Phi^1(t) &= \int_0^t T_{t-s_1}BT_{s_1}\, dW(s_1), \end{aligned}\right\}$$

for $t \geq 0$.

First, we show by induction that for each $n \geq 1$, the process $\Phi^n : [0,\infty)\times\Omega \to L_2(H)$ has a version with a.a. sample paths continuous on $[0,\infty)$. In view of (1.2.21), this will follow from Proposition (7.3) ([D-Z.1], p. 184) provided we show that

(1.2.22)
$$\int_0^a E\|\Phi^{n-1}(t)\|^{2p}_{L_2(H)}\, dt < \infty$$

for all integers $n > 1$ and $p \in (1, \alpha^{-1}]$. For later use, we will actually prove the stronger estimate

(1.2.22′)
$$E \sup_{0\leq s\leq t} \|\Phi^n(s)\|^{2p}_{L_2(H)} \leq K_1 \frac{(K_2 t)^{n-1}}{(n-1)!}, \quad t \in [0,a],$$

for all integers $n \geq 1$, and $p \in (1, \alpha^{-1}]$, where K_1, K_2 are positive constants depending only on p and a. We use induction on n to establish (1.2.22′). To check

(1.2.22′) for $n = 1$, choose $p \in (1, \alpha^{-1}]$, and consider the following easy estimates:

$$\begin{aligned}\left\{\int_0^a \|T_s\|_{L_2(H)}^{2p}\, ds\right\}^{1/p} &= \left\{\int_0^a \left[\sum_{k=1}^\infty e^{-2\mu_k s}\right]^p ds\right\}^{1/p}\\ &\leq \sum_{k=1}^\infty \left\{\int_0^a e^{-2\mu_k p s}\, ds\right\}^{1/p}\\ &\leq \frac{1}{(2p)^{1/p}} \sum_{k=1}^\infty \mu_k^{-1/p}\\ &\leq \frac{1}{(2p)^{1/p}} \sum_{k=1}^\infty \mu_k^{-\alpha} < \infty.\end{aligned}$$

Now use the second equality in (1.2.21) and Proposition (7.3) ([D-Z.1], p. 184) to get the following estimate:

$$E \sup_{0\leq s\leq t} \|\Phi^1(s)\|_{L_2(H)}^{2p} \leq C_1 \int_0^t \|T_{s_1}\|_{L_2(H)}^{2p}\, ds_1 \leq \frac{C_1}{2p}\left[\sum_{k=1}^\infty \mu_k^{-\alpha}\right]^p \tag{1.2.23}$$

for all $t \in [0, a]$ and for $p \in (1, \alpha^{-1}]$. The constant C_1 does not depend on $t \in [0, a]$. Since $A^{-\alpha}$ is trace-class, the above inequality implies that (1.2.22′) holds with $K_1 := \frac{C_1}{2p}\left[\sum_{k=1}^\infty \mu_k^{-\alpha}\right]^p$, for $n = 1$, and any $p \in (1, \alpha^{-1}]$. Now suppose that (1.2.22′) holds for some integer $n \geq 1$ and all $p \in (1, \alpha^{-1}]$. Then the first equality in (1.2.21) and Proposition (7.3) ([D-Z.1], p. 184) imply that there is a positive constant $K_2 := K_2(p, a)$ such that

$$\begin{aligned}E \sup_{0\leq s\leq t} \|\Phi^{n+1}(s)\|_{L_2(H)}^{2p} &\leq K_2 \int_0^t E\|\Phi^n(s_1)\|_{L_2(H)}^{2p}\, ds_1\\ &\leq K_2 \int_0^t K_1 \frac{(K_2 s_1)^{n-1}}{(n-1)!}\, ds_1 = K_1 \frac{(K_2 t)^n}{n!},\end{aligned} \tag{1.2.24}$$

for all $t \in [0, a]$ and $p \in (1, \alpha^{-1}]$. Therefore by induction, (1.2.22′) (and hence (1.2.22)) hold for all integers $n \geq 1$ and any $p \in (1, \alpha^{-1}]$.

From the first equality in (1.2.21), (1.2.22) and Proposition 7.3 ([D-Z.1], p. 184), it follows that each $\Phi^n : [0, \infty) \times \Omega \to L_2(H)$ has a version with a.a. sample paths continuous on $[0, \infty)$. From the estimate (1.2.22′), it is easy to see that the series $\sum_{n=1}^\infty \Phi^n$ converges absolutely in $L^{2p}(\Omega, C([0, a], L_2(H)))$ for each $a > 0$ and $p \in (1, \alpha^{-1}]$. This gives a continuous modification for the sum $\sum_{n=1}^\infty \Phi^n$ of the series in (1.2.18). Hence the $L_2(H)$-valued process

$$u(t, \cdot, \cdot) - T_t = \sum_{n=1}^\infty \Phi^n(t), \quad t \geq 0,$$

has a version with almost all sample-paths continuous on $[0, \infty)$. This proves the first assertion in (iv). To prove the second assertion in (iv), it suffices to show that the mapping $(0,\infty) \ni t \mapsto T_t \in L_2(H)$ is locally Lipschitz. To see this, let $0 < t_1 < t_2 \le a < \infty$. Then

$$\begin{aligned}\|T_{t_2} - T_{t_1}\|^2_{L_2(H)} &\le \sum_{k=1}^{\infty} [e^{-\mu_k t_2} - e^{-\mu_k t_1}]^2 \\ &\le (t_2 - t_1)^2 \sum_{k=1}^{\infty} \mu_k^2 e^{-2\mu_k t_1} \\ &\le \frac{3}{4t_1^3} (t_2 - t_1)^2 \sum_{k=1}^{\infty} \mu_k^{-1}.\end{aligned}$$

Since A^{-1} is trace-class, the above inequality implies that the mapping $(0,\infty) \ni t \mapsto T_t \in L_2(H)$ is locally Lipschitz. The second assertion in (iv) now follows immediately from this and the first assertion.

The measurability assertions in (iv) follow directly from the relation

$$u(t,\cdot,\cdot) = T_t + \sum_{n=1}^{\infty} \Phi^n(t), \quad t > 0,$$

and the fact that, as $L_2(H)$-valued Itô stochastic integrals, the processes $\Phi^n : (0,\infty) \times \Omega \to L_2(H)$, $n \ge 1$, are $(\mathcal{F}_t)_{t>0}$-adapted and $(\mathcal{B}((0,\infty)) \otimes \mathcal{F}, \mathcal{B}(L_2(H)))$-measurable.

The evaluation map

$$\begin{aligned} L_2(H) \times H &\to H \\ (S, x) &\mapsto S(x) \end{aligned}$$

is continuous bilinear. Therefore the first assertion in (iv) implies that the map $[0,T] \times H \ni (t,x) \to u(t,x,\omega) - T_t(x) \in H$ is jointly continuous for almost all $\omega \in \Omega$. Since $[0,T] \times H \ni (t,x) \to T_t(x) \in H$ is jointly continuous (by strong continuity of the semigroup $T_t, t \ge 0$), the first assertion in (ii) follows.

Finally, it remains to prove the estimate in (ii). In view of (1.2.22′), the series in (1.2.18) converges absolutely in $L^{2p}(\Omega, C([0,a], L_2(H)))$, $p \in (1, \alpha^{-1}]$. Therefore,

$$\begin{aligned}&\{E \sup_{0\le t\le a} \|u(t,\cdot,\cdot)\|^{2p}_{L(H)}\}^{1/(2p)} \\ &\quad\le \sup_{0\le t\le a} \|T_t\|_{L(H)} + \sum_{n=1}^{\infty} \{E \sup_{0\le t\le a} \|\Phi^n(t)\|^{2p}_{L_2(H)}\}^{1/(2p)} \\ &\quad\le \sup_{0\le t\le a} \|T_t\|_{L(H)} + K_1^{1/(2p)} \sum_{n=1}^{\infty} \left\{\frac{(K_2 a)^{n-1}}{(n-1)!}\right\}^{1/(2p)} < \infty.\end{aligned}$$

This proves the estimate in (ii), and the proof of Theorem 1.2.1 is complete. □

THEOREM 1.2.2. *Assume the following:*

(i) A^{-1} *is a trace class operator, i.e.,* $\sum_{n=1}^{\infty} \mu_n^{-1} < \infty$.

(ii) $T_t \in L(H)$, $t \ge 0$, *is a strongly continuous contraction semigroup.*

Then the mild solution of the linear stochastic evolution equation (1.2.1) *has a version* $u : \mathbf{R}^+ \times H \times \Omega \to H$ *which satisfies the assertions* (i), (iii) *and* (iv) *of*

Theorem 1.2.1. *Furthermore, for almost all $\omega \in \Omega$, the map $[0,\infty) \times H \ni (t,x) \to u(t,x,\omega) \in H$ is jointly continuous, and for any fixed $a \in \mathbf{R}^+$,*

$$E \sup_{0 \le t \le a} \|u(t,\cdot,\cdot)\|^2_{L(H)} < \infty.$$

PROOF. The proof follows that of Theorem 1.2.1. We will only highlight the differences.

We assume Hypotheses (i) and (ii). By the proof of Theorem 1.2.1, Hypothesis (i) implies that the solution of (1.2.1) admits a version $u : \mathbf{R}^+ \times H \times \Omega \to H$ which satisfies assertions (i) and (iii) of Theorem 1.2.1.

Use the notation in the proof of Theorem 1.2.1. In particular, one has

$$u(t,\cdot,\cdot) - T_t = \sum_{n=1}^{\infty} \Phi^n(t), \quad t \ge 0,$$

where the series converges in $L^2(\Omega, L_2(H))$ for each $t \ge 0$. Fix any $a > 0$. Since A^{-1} is trace-class, then

$$\int_0^a \|T_s\|^2_{L_2(H)}\, ds < \infty.$$

Using this, the fact that T_t, $t \ge 0$, is a contraction semigroup, and Theorem 6.10 ([D-Z.1], p. 160), it follows that $\Phi^1 : [0,\infty) \times \Omega \to L_2(H)$ has a sample-continuous version. Furthermore, there is a positive constant K_3 such that

$$E \sup_{0 \le s \le t} \|\Phi^1(s)\|^2_{L_2(H)} \le K_3 \int_0^t \|T_s\|^2_{L_2(H)}\, ds < \infty, \tag{1.2.25}$$

for all $t \in [0,a]$ ([D-Z.1], Theorem 6.10, p. 160). We will show that the series $\sum_{n=2}^{\infty} \Phi^n$ converges in $L^{2p}(\Omega, C([0,a], L_2(H))$ for all $p \ge 1$. Therefore, the series $\sum_{n=1}^{\infty} \Phi^n$ converges in $L^2(\Omega, C([0,a], L_2(H)))$.

By Lemma 7.2, ([D-Z.1], p. 182), we have

$$E\|\Phi^1(t)\|^{2p}_{L_2(H)} \le K_4 \left[\int_0^t \|T_s\|^2_{L_2(H)}\, ds\right]^p \tag{1.2.26}$$

for all $t \in [0,a]$ and all $p \ge 1$. The constant K_4 depends on p but is independent of $t \in [0,a]$. Since A^{-1} is trace-class, the above inequality, Proposition 7.3 ([D-Z.1], p. 184) and an induction argument imply the following inequality:

$$E \sup_{0 \le s \le t} \|\Phi^n(s)\|^{2p}_{L_2(H)} \le K_5 \frac{(K_6 t)^{n-1}}{(n-1)!}, \quad t \in [0,a],$$

for all integers $n \ge 2$, and $p \ge 1$, where K_5, K_6 are positive constants depending only on p and a (cf. (1.2.22′) in the proof of Theorem 1.2.1). The rest of the proof of the theorem follows from the above inequality by a similar argument to the one in the proof of Theorem 1.2.1. □

THEOREM 1.2.3. *Assume that* $\sum_{n=1}^{\infty} \mu_n^{-1} \|B(e_n)\|^2_{L_2(K,H)} < \infty$.

Then the mild solution of the linear stochastic evolution equation (1.2.1) *has a* $(\mathcal{B}(\mathbf{R}^+) \otimes \mathcal{B}(H) \otimes \mathcal{F}, \mathcal{B}(H))$*-measurable version* $u : \mathbf{R}^+ \times H \times \Omega \to H$ *with the following properties:*

(i) *For each* $x \in H$, *the process* $u(\cdot, x, \cdot) : \mathbf{R}^+ \times \Omega \to H$ *is* $(\mathcal{B}(\mathbf{R}^+) \otimes \mathcal{F}, \mathcal{B}(H))$*-measurable,* $(\mathcal{F}_t)_{t \geq 0}$*-adapted and satisfies the stochastic integral equation* (1.2.2).

(ii) *For almost all* $\omega \in \Omega$, *the map* $[0, \infty) \times H \ni (t, x) \to u(t, x, \omega) \in H$ *is jointly continuous. Furthermore, for any fixed* $a \in \mathbf{R}^+$,

$$E \sup_{0 \leq t \leq a} \|u(t, \cdot, \cdot)\|^2_{L(H)} < \infty.$$

(iii) *For almost all* $\omega \in \Omega$ *and each* $t > 0$, $u(t, \cdot, \omega) : H \to H$ *is a bounded linear operator with the following representation:*

$$u(t, \cdot, \cdot) = T_t + \sum_{n=1}^{\infty} \int_0^t T_{t-s_1} B \int_0^{s_1} T_{s_1 - s_2} B \cdots$$
$$\cdots \int_0^{s_{n-1}} T_{s_{n-1}-s_n} B T_{s_n} \, dW(s_n) \cdots dW(s_2) \, dW(s_1).$$

In the above equation, the iterated Itô stochastic integrals are interpreted in the sense of Lemma 1.2.2, *and the convergence of the series holds in the Hilbert space* $L_2(H)$ *of Hilbert-Schmidt operators on* H. *If in addition,* $T_t : H \to H$ *is compact for each* $t > 0$, *then so is* $u(t, \cdot, \omega) : H \to H$ *for almost all* $\omega \in \Omega$.

(iv) *For almost all* $\omega \in \Omega$, *the path* $[0, \infty) \ni t \mapsto u(t, \cdot, \omega) - T_t \in L_2(H)$ *is continuous. Furthermore, the process* $u : (0, \infty) \times \Omega \to L(H)$ *is* $(\mathcal{F}_t)_{t \geq 0}$*-adapted and* $(\mathcal{B}((0, \infty)) \otimes \mathcal{F}, \mathcal{B}(L(H)))$*-measurable.*

PROOF. The proof follows along the same lines as that of Theorem 1.2.1. Just observe that the hypothesis of Theorem 1.2.3 implies the following integrability property

$$\int_0^a \|BT_t\|^2_{L_2(H, L_2(K,H))} \, dt < \infty$$

for any $a > 0$. □

REMARK. It is easy to see that the hypothesis of Theorem 1.2.3 is satisfied if one assumes that the mapping $B : H \to L_2(K, H)$ is Hilbert-Schmidt. By contrast to the hypotheses of Theorems 1.2.1, 1.2.2, the assumption in Theorem 1.2.3 does not entail any dimension restriction if the operator A is a differential operator on a Euclidean domain. Furthermore, one does not require even discreteness of the spectrum of A if we assume that $B : H \to L_2(K, H)$ is Hilbert-Schmidt. However, in this case, one gets a flow of *bounded linear* (but not necessarily compact) maps $u(t, \cdot, \omega) \in L(H)$, $t > 0, \omega \in \Omega$.

We will continue to assume the hypotheses of Theorem 1.2.1, 1.2.2 or 1.2.3.

Let $u : \mathbf{R}^+ \times \Omega \to L(H)$ be the regular version of the mild solution of (1.2.1) given by Theorems 1.2.1, 1.2.2 or 1.2.3. Our next result in this section identifies u as a *fundamental solution* (or *parametrix*) for (1.2.1).

Consider the following stochastic integral equation:

$$\left.\begin{aligned} v(t) &= T_t + \int_0^t T_{t-s}Bv(s)dW(s), \quad t>0 \\ v(0) &= I, \end{aligned}\right\} \tag{1.2.27}$$

where I denotes the identity operator on H and the stochastic integral is interpreted as an Itô integral in the Hilbert space $L_2(H)$.

REMARK. The initial-value problem (1.2.27) cannot be interpreted strictly in the Hilbert space $L_2(H)$ since $v(0) = I \notin L_2(H)$. On the other hand, one cannot view the equation (1.2.27) in the Banach space $L(H)$, because the latter Banach space is not sufficiently "smooth" to allow for a satisfactory theory of stochastic integration.

We say that a stochastic process $v : [0,\infty) \times \Omega \to L(H)$ is a *solution* to equation (1.2.27) if

(i) $v : (0,\infty) \times \Omega \to L_2(H)$ is $(\mathcal{F}_t)_{t>0}$-adapted, and $(\mathcal{B}((0,\infty)) \otimes \mathcal{F}, \mathcal{B}(L_2(H)))$-measurable.

(ii) $v \in L^2((0,a) \times \Omega, L_2(H))$ for all $a \in (0,\infty)$.

(iii) v satisfies (1.2.27) almost surely.

THEOREM 1.2.3′. *Assume the hypotheses of* Theorem 1.2.1, 1.2.2 or 1.2.3. *Let u be the regular version of the mild solution of* (1.2.1) *given therein. Then u is the unique solution of* (1.2.27) *in* $L^2((0,a) \times \Omega, L_2(H))$ *for* $a > 0$.

PROOF. Assume the hypotheses of Theorems 1.2.1, 1.2.2 or 1.2.3. Let u be the regular version of the mild solution of (1.2.1) given by these theorems.

Note first that $u : (0,\infty) \times \Omega \to L_2(H)$ is $(\mathcal{F}_t)_{t\geq 0}$-adapted, and $(\mathcal{B}((0,\infty)) \otimes \mathcal{F}, \mathcal{B}(L_2(H)))$-measurable. This follows from assertion (iv) in Theorem 1.2.1.

In the proofs of Theorems 1.2.1, 1.2.2, we have shown that the series $\sum_{n=1}^{\infty} \Phi^n$ converges absolutely in $L^2(\Omega, C([0,a], L_2(H)))$, and hence also in the Hilbert space $L^2(\Omega, L^2((0,a), L_2(H)))$, because of the continuous linear embedding

$$L^2(\Omega, C([0,a], L_2(H))) \subset L^2(\Omega, L^2((0,a), L_2(H))) \equiv L^2((0,a) \times \Omega, L_2(H)).$$

Thus

$$\begin{aligned} &\int_0^a E\|u(t,\cdot)\|^2_{L_2(H)}\, dt \\ &\qquad \leq 2\int_0^a \|T_t\|^2_{L_2(H)}\, dt + 2\left[\sum_{n=1}^{\infty}\left\{\int_0^a E\|\Phi^n(t)\|^2_{L_2(H)}\, dt\right\}^{1/2}\right]^2 < \infty. \end{aligned}$$

In particular, the Itô stochastic integral $\int_0^t T_{t-s}Bu(s)\, dW(s)$ is well-defined in $L_2(H)$ for each $t \in [0,a]$ (Lemma 1.2.2).

We next show that u solves the operator-valued stochastic integral equation (1.2.27). To see this, use the fact that

$$\left[\int_0^t T_{t-s}Bu(s,\cdot)\, dW(s)\right](e_n) = \int_0^t T_{t-s}Bu(s,e_n,\cdot)\, dW(s), \quad n \geq 1,$$

and the integral equation (1.2.2) to conclude that

$$(1.2.28)\quad u(t,\omega)(e_n) = T_t e_n + \left[\int_0^t T_{t-s}Bu(s,\cdot,\cdot)\,dW(s)\right](\omega)(e_n), \quad t \geq 0,\, n \geq 1,$$

holds for *all* ω in a sure event $\Omega^* \in \mathcal{F}$ which is independent of n and $t \geq 0$. Since $\{e_n : n \geq 1\}$ is a complete orthonormal system in H, it follows from (1.2.28) that for all $\omega \in \Omega^*$, one has

$$(1.2.29)\qquad u(t,\omega)(x) = T_t x + \left[\int_0^t T_{t-s}Bu(s,\cdot,\cdot)\,dW(s)\right](\omega)(x), \quad t \geq 0,\, n \geq 1$$

for all $x \in H$. Thus u is a solution of (1.2.27).

Finally we show that (1.2.27) has a unique $(\mathcal{F}_t)_{t>0}$-adapted solution in $L^2((0,a) \times\Omega, L_2(H))$. Suppose v_1, v_2 are two such solutions of (1.2.27). Then

(1.2.30)

$$E\|v_1(t) - v_2(t)\|^2_{L_2(H)} \leq \|B\|_{L_2(K,H)} \sup_{0\leq u\leq a} \|T_u\|_{L(H)} \int_0^t E\|v_1(s) - v_2(s)\|^2_{L_2(H)}\,ds$$

for all $t \in (0, a]$. The above inequality implies that $E\|v_1(t) - v_2(t)\|^2_{L_2(H)} = 0$ for all $t > 0$ and uniqueness holds. □

From now on and throughout this section, we will impose the following

Condition (B):

(i) The operator $B : H \to L_2(K, H)$ can be extended to a bounded linear operator $H \to L(E, H)$, which will also be denoted by B.

(ii) The series $\sum_{k=1}^{\infty} \|B_k^2\|_{L(H)}$ converges, where the bounded linear operators $B_k : H \to H$ are defined by $B_k(x) := B(x)(f_k), x \in H,\, k \geq 1$.

THEOREM 1.2.4. *Assume the hypotheses of* Theorem 1.2.1, 1.2.2 *or* 1.2.3, *and* Condition (B) *are satisfied. Then the mild solution of* (1.2.1) *admits a version* $u : \mathbf{R}^+ \times \Omega \to L(H)$ *satisfying the conclusions of* Theorem 1.2.1, 1.2.2 *or* 1.2.3 *and is such that*

(i) (u, θ) *is a perfect* $L(H)$*-valued cocycle:*

$$(1.2.31)\qquad u(t + s, \omega) = u(t, \theta(s, \omega)) \circ u(s, \omega)$$

for all $s, t \geq 0$ *and all* $\omega \in \Omega$;

(ii) $\sup_{0\leq s\leq t\leq a} \|u(t - s, \theta(s, \omega))\|_{L(H)} < \infty$, *for all* $\omega \in \Omega$ *and all* $a > 0$.

PROOF. In view of Theorem 1.2.3′, u satisfies the stochastic integral equation

$$(1.2.32)\qquad \left.\begin{aligned} u(t) &= T_t + \int_0^t T_{t-s}Bu(s)\,dW(s), \quad t > 0 \\ u(0) &= I \end{aligned}\right\}$$

with $u(t) \in L_2(H)$ a.s. for all $t > 0$.

Our strategy for proving the cocycle property (1.2.31) is to approximate the cylindrical Wiener process W in (1.2.32) by a suitably defined family of smooth processes $W_n : \mathbf{R}^+ \times \Omega \to E$, $n \geq 1$, prove the cocycle property for the corresponding approximating solutions and then pass to the limit in $L_2(H)$ as n tends to ∞.

Define W_n on $\mathbf{R}^+ \times \Omega$, $n \geq 1$, by

$$W_n(t,\omega) := n\int_{t-1/n}^{t} W(u,\omega)\,du - n\int_{-1/n}^{0} W(u,\omega)\,du, \quad t \geq 0,\ \omega \in \Omega. \tag{1.2.33}$$

It is easy to see that each W_n is a helix:

$$W_n(t,\theta(t_1,\omega)) = W_n(t+t_1,\omega) - W_n(t_1,\omega), \tag{1.2.34}$$

and

$$W_n'(t,\theta(t_1,\omega)) = W_n'(t+t_1,\omega) \tag{1.2.35}$$

for all $t, t_1 \geq 0$, $\omega \in \Omega$, $n \geq 1$. In (1.2.35), the prime $'$ denotes differentiation with respect to t.

For each $k \geq 1$, recall the definition of $B_k : H \to H$ in Condition (B)(ii). For each integer $n \geq 1$, define the process $u_n : \mathbf{R}^+ \times \Omega \to L_2(H)$ to be the unique $(\mathcal{B}((0,\infty)) \otimes \mathcal{F}, \mathcal{B}(L_2(H)))$-measurable, $(\mathcal{F}_t)_{t>0}$-adapted solution of the random integral equation:

$$\left.\begin{aligned} u_n(t,\omega) =& T_t + \int_0^t T_{t-s}\circ\{[B\star u_n(s,\omega)](W_n'(s,\omega))\}\,ds \\ & \qquad - \frac{1}{2}\int_0^t \sum_{k=1}^{\infty} T_{t-s}\circ B_k^2 \circ u_n(s,\omega)\,ds, \quad t>0 \\ u_n(0,\omega) =& I, \end{aligned}\right\} \tag{1.2.27)(n}$$

for $\omega \in \Omega$. Recall that the operation $\star$ is defined by (1.2.16) in the proof of Lemma 1.2.2.

Then

$$\lim_{n\to\infty} \sup_{0<t\leq a} \|u_n(t) - u(t)\|^2_{L_2(H)} = 0, \tag{1.2.36}$$

in probability, for each $a > 0$. The convergence (1.2.36) follows by modifying the proof (in $L_2(H)$) of the Wong-Zakai approximation theorem for stochastic evolution equations in ([Tw], Theorem 1.3.4.1). (Cf. [I-W], Theorem 7.2, p. 497).

Next, we show that for each $n \geq 1$, (u_n, θ) is a perfect cocycle. Fix $n \geq 1$, $t_1 \geq 0$ and $\omega \in \Omega$. Using (1.2.27)(n), it follows that

$$\begin{aligned} & u_n(t,\theta(t_1,\omega)) \circ u_n(t_1,\omega) \\ & = T_t \circ u_n(t_1,\omega) \\ & + \int_{t_1}^{t+t_1} T_{t+t_1-s}\circ\{[B\star(u_n(s-t_1,\theta(t_1,\omega))\circ u_n(t_1,\omega))](W_n'(s-t_1,\theta(t_1,\omega)))\}\,ds \\ & - \frac{1}{2}\int_{t_1}^{t+t_1} \sum_{k=1}^{\infty} T_{t+t_1-s}\circ B_k^2 \circ (u_n(s-t_1,\theta(t_1,\omega))\circ u_n(t_1,\omega))\,ds \\ & = T_{t+t_1} + \int_0^{t_1} T_{t+t_1-s}\circ\{[B\star u_n(s,\omega)](W_n'(s,\omega))\}\,ds \end{aligned}$$

$$-\frac{1}{2}\int_0^{t_1}\sum_{k=1}^{\infty} T_{t+t_1-s}\circ B_k^2\circ u_n(s,\omega)\,ds$$
$$+\int_{t_1}^{t+t_1} T_{t+t_1-s}\circ\{[B\star(u_n(s-t_1,\theta(t_1,\omega))\circ u_n(t_1,\omega))](W_n'(s-t_1,\theta(t_1,\omega)))\}\,ds$$
$$-\frac{1}{2}\int_{t_1}^{t+t_1}\sum_{k=1}^{\infty} T_{t+t_1-s}\circ B_k^2\circ(u_n(s-t_1,\theta(t_1,\omega))\circ u_n(t_1,\omega))\,ds,$$

for $t>0$. Hence, using (1.2.27)(n) and (1.2.35), we obtain

$$\begin{aligned}
u_n(t,&\theta(t_1,\omega))\circ u_n(t_1,\omega)-u_n(t_1+t,\omega)\\
&=\int_{t_1}^{t+t_1} T_{t+t_1-s}\circ\{[B\star(u_n(s-t_1,\theta(t_1,\omega))\circ u_n(t_1,\omega)-u_n(s,\omega))]\\
&\qquad\qquad (W_n'(s-t_1,\theta(t_1,\omega)))\}\,ds\\
&\quad-\frac{1}{2}\int_{t_1}^{t+t_1}\sum_{k=1}^{\infty} T_{t+t_1-s}\circ B_k^2\circ[u_n(s-t_1,\theta(t_1,\omega))\circ u_n(t_1,\omega)-u_n(s,\omega)]\,ds\\
&=\int_0^{t} T_{t-s}\circ\{[B\star(u_n(s,\theta(t_1,\omega))\circ u_n(t_1,\omega)-u_n(s+t_1,\omega))](W_n'(s,\omega))\}\,ds\\
&\quad-\frac{1}{2}\int_0^{t}\sum_{k=1}^{\infty} T_{t-s}\circ B_k^2\circ[u_n(s,\theta(t_1,\omega))\circ u_n(t_1,\omega)-u_n(s+t_1,\omega)]\,ds,
\end{aligned}$$

for all $t>0$. The above identity and a simple application of Gronwall's lemma yields

$$u_n(t,\theta(t_1,\omega))\circ u_n(t_1,\omega)-u_n(t_1+t,\omega)=0 \tag{1.2.37}$$

for all $t,t_1\geq 0$ and all $\omega\in\Omega$. Hence (u_n,θ) is a perfect cocycle in $L(H)$. Using (1.2.36) and passing to the limit in $L(H)$ as $n\to\infty$ in the above identity implies that (u,θ) is a crude $L(H)$-valued cocycle. In order to obtain a perfect version of this cocycle, it is sufficient to prove that there is a sure event $\Omega^*\in\mathcal{F}$ (independent of $t_1\in\mathbf{R}^+$) such that $\theta(t,\cdot)(\Omega^*)\subseteq\Omega^*$ for all $t\geq 0$, and there is a subsequence $\{u_{n'}\}_{n'=1}^{\infty}$ of $\{u_n\}_{n=1}^{\infty}$ such that

$$\lim_{n',m'\to\infty}\sup_{0<t\leq a}\|u_{n'}(t,\omega)-u_{m'}(t,\omega)\|_{L_2(H)}^2=0, \tag{1.2.38}$$

for each $a>0$ and all $\omega\in\Omega^*$. Set $v_n(t_1,t,\omega):=u_n(t-t_1,\theta(t_1,\omega))$, $t\geq t_1\geq 0$. Then v_n solves the integral equation

$$\begin{aligned}
v_n(t_1,t,\omega)=&T_{t-t_1}+\int_{t_1}^{t} T_{t-s}\circ\{[B\star v_n(t_1,s,\omega)](W_n'(s,\omega))\}\,ds\\
&-\frac{1}{2}\int_{t_1}^{t}\sum_{k=1}^{\infty} T_{t-s}\circ B_k^2\circ v_n(t_1,s,\omega)\,ds,\\
v_n(t_1,t_1,\omega)=&I,
\end{aligned}$$

for $t\geq t_1\geq 0$. The above equation implies that $v_n(t_1,t,\omega)$ is continuous in (t_1,t) for each $\omega\in\Omega$. Furthermore, if we apply the approximation scheme (in $L_2(H)$) to the above integral equation, we get a subsequence $\{v_{n'}\}_{n'=1}^{\infty}$ of $\{v_n\}_{n=1}^{\infty}$ such that for a.a. $\omega \in \Omega$

$$\lim_{n',m'\to\infty} \sup_{0<t_1\leq t\leq a} \|v_{n'}(t_1,t,\omega) - v_{m'}(t_1,t,\omega)\|_{L_2(H)}^2 = 0, \tag{1.2.39}$$

for each $a > 0$. Now define Ω^* to be the set of all $\omega \in \Omega$ such that the subsequence $\{v_{n'}(t_1,t,\omega) : n' \geq 1\}$ converges in $L(H)$ uniformly in (t_1,t) for $0 < t_1 \leq t \leq a$ and all $a > 0$. Therefore Ω^* is a $\theta(t,\cdot)$-invariant sure event. Define

$$u(t,\omega) := \lim_{n'\to\infty} v_{n'}(0,t,\omega)$$

for all $t \geq 0$ and all $\omega \in \Omega^*$. Hence (u,θ) is a perfect cocycle in $L(H)$. This proves assertion (i) of the theorem.

To prove the second assertion of the theorem, fix $s \geq 0$ and define $\hat{v}_n(s,t,\omega) := \hat{u}_n(t-s,\theta(s,\omega)) = u_n(t-s,\theta(s,\omega)) - T_{t-s}$, $t \geq s \geq 0$. It is easy to see that $\hat{v}_n$ solves the integral equation

$$\begin{aligned}
\hat{v}_n(s,t,\omega) = &\int_s^t T_{t-\lambda}\circ\{[B\star\hat{v}_n(s,\lambda,\omega)](W_n'(\lambda,\omega))\}\,d\lambda \\
&+\int_s^t T_{t-\lambda}BT_{\lambda-s}(W_n'(\lambda,\omega))\,d\lambda \\
&-\frac{1}{2}\int_s^t\sum_{k=1}^{\infty} T_{t-\lambda}\circ B_k^2\circ\hat{v}_n(s,\lambda,\omega)\,d\lambda, \\
\hat{v}_n(s,s,\omega) =& 0 \in L_2(H),
\end{aligned}$$

for $t \geq s \geq 0$. The above equation implies that the map $\Delta \ni (s,t) \mapsto \hat{v}_n(s,t,\omega) \in L_2(H)$ is continuous for each $\omega \in \Omega$. Applying the approximation scheme again, there is a subsequence $\{\hat{v}_{n'}\}_{n'=1}^{\infty}$ of $\{\hat{v}_n\}_{n=1}^{\infty}$ such that for a.a. $\omega \in \Omega$, one has

$$\lim_{n',m'\to\infty} \sup_{0\leq s\leq t\leq a} \|\hat{v}_{n'}(s,t,\omega) - \hat{v}_{m'}(s,t,\omega)\|_{L_2(H)}^2 = 0,$$

for each $a > 0$. Define $\hat{\Omega}^*$ to be the set of all $\omega \in \Omega$ such that the subsequence $\{\hat{v}_{n'}(s,t,\omega) : n' \geq 1\}$ converges in $L_2(H)$ uniformly in (s,t) for $0 \leq s \leq t \leq a$ and all $a > 0$. Therefore $\hat{\Omega}^*$ is a $\theta(t,\cdot)$-invariant sure event. Define

$$\hat{u}(t,\omega) := \lim_{n'\to\infty} \hat{v}_{n'}(0,t,\omega)$$

for all $t \geq 0$ and all $\omega \in \hat{\Omega}^*$. Therefore, the map $\Delta \ni (s,t) \mapsto \hat{u}(t-s,\theta(s,\omega)) \in L_2(H)$ is jointly continuous. In particular, $\sup_{0\leq s\leq t\leq a} \|\hat{u}(t-s,\cdot,\theta(s,\omega))\|_{L(H)} < \infty$, for all $\omega \in \hat{\Omega}^*$ and all $a > 0$. Using the fact that $\sup_{0\leq s\leq t\leq a} \|T_{t-s}\|_{L(H)} < \infty$, it follows that $u(t,\omega) := \hat{u}(t,\omega) + T_t$, $\omega \in \Omega^* \cap \hat{\Omega}^*$, gives a version of the cocycle that also satisfies assertion (ii) of the theorem. This completes the proof of the theorem. □

REMARKS.

(i) Results analogous to Theorem 1.2.4 hold if B is replaced by the an affine linear map $B(x) := B_0 + B_1(x)$, $x \in H$, where $B_0 \in L(E,H)$ and $B_1 : H \to L(E,H)$ satisfies Condition (B). In this case, one gets a cocycle (u,θ) where each map $u(t,\cdot,\omega) : H \to H$ is of the form $u(t,\cdot,\omega) = u_0(t,\cdot,\omega) + u_1(t,\omega)$ with

$u_0(t,\cdot,\omega) \in L_2(H)$ and $u_1(t,\omega) \in H$ for $t > 0, \omega \in \Omega$. This follows using minor modifications of the above arguments.

(ii) It is possible to replace B in the see (1.2.1) by an adapted random field $B : \mathbf{R}^+ \times H \times \Omega \to L(E,H)$ satisfying appropriate integrability and regularity conditions, and which is such that $B(t,\cdot,\omega) : H \to L(E,H)$ satisfies Condition (B) for each $t \geq 0$, $\omega \in \Omega$. The conclusions of Theorems 1.2.1-1.2.3, 1.2.3′ will still hold in this case. However, the stochastic semiflow will only satisfy Definition 1.1.1 (rather than the cocycle property in Definition 1.1.2). On the other hand if B is stationary, then the cocycle property should hold (on a suitably enlarged probability space) (Theorem 1.2.4).

(iii) Theorems 1.2.1-1.2.4, 1.2.3′ also hold if the operator A is allowed to have a non-zero discrete spectrum $\{\mu_n : n \geq 1\}$ which is bounded below. This yields a splitting $A = A_0 + A_1$ where $\sigma(A_0)$ consists of positive eigenvalues and $\sigma(A_1)$ of finitely many negative eigenvalues.

(b) *Semilinear see's*

In this section, we continue to assume that the operators A, B, the cylindrical Brownian motion W, the complete filtered Wiener space $(\Omega, \bar{\mathcal{F}}, (\mathcal{F}_t)_{t\geq 0}, P)$ and the Brownian shift $\theta : \mathbf{R} \times \Omega \to \Omega$ are as defined in part (a) of this section and satisfy the conditions therein. The semigroup generated by $-A$ is denoted as before by $T_t, t \geq 0$. Furthermore, we let $F : H \to H$ be a (Fréchet) C^1 non-linear map satisfying the following locally Lipschitz and linear growth hypotheses:

$$
\left.\begin{aligned}
|F(v)| &\leq C(1+|v|), \quad v \in H \\
|F(v_1) - F(v_2)| &\leq L_n |v_1 - v_2|, \quad v_i \in H, |v_i| \leq n, i = 1,2,
\end{aligned}\right\} \tag{1.2.40}
$$

for some positive constants $C, L_n, n \geq 1$.

Consider the semilinear see:

$$
\left.\begin{aligned}
du(t) &= -Au(t)dt + F(u(t))dt + Bu(t)\, dW(t), \quad t > 0, \\
u(0) &= x \in H,
\end{aligned}\right\} \tag{1.2.41}
$$

where the operators A, B satisfy the hypotheses of Theorem 1.2.4.

Our main objective in this section is to establish the existence of a C^k perfect cocycle (U, θ) for the above stochastic evolution equation. First we define a *mild solution* of (1.2.41) as a family of $(\mathcal{B}(\mathbf{R}^+) \otimes \mathcal{F}, \mathcal{B}(H))$-measurable, $(\mathcal{F}_t)_{t\geq 0}$-adapted processes $u(\cdot, x, \cdot) : \mathbf{R}^+ \times \Omega \to H$, $x \in H$, satisfying the following stochastic integral equations:

$$
u(t,x,\cdot) = T_t(x) + \int_0^t T_{t-s}(F(u(s,x,\cdot)))\, ds + \int_0^t T_{t-s} Bu(s,x,\cdot)\, dW(s), \tag{1.2.42}
$$

for $t \geq 0$, a.s. ([D-Z.1], Chapter 7, p. 182).

To fix notation, denote by $\phi : \mathbf{R}^+ \times \Omega \to L(H)$ the perfect cocycle generated by the linear stochastic evolution equation

$$
\left.\begin{aligned}
d\phi(t) &= -\, A\phi(t)dt + B\phi(t)\, dW(t), \quad t > 0, \\
\phi(0) &= I \in L(H),
\end{aligned}\right\} \tag{1.2.43}
$$

and obtained via Theorem 1.2.4. In particular, $\phi(t,\omega) \in L_2(H), t > 0, \omega \in \Omega$, stands for $u(t,\cdot,\omega)$ in the notation of part (a) of this section.

Our first step in the construction of a non-linear cocycle of (1.2.41) is to observe that mild solutions of (1.2.41) correspond to solutions of a *random* integral equation on H. This is shown in the following theorem:

THEOREM 1.2.5. *Suppose the hypotheses of* Theorem 1.2.4 *are satisfied. Then every* $(\mathcal{B}(\mathbf{R}^+)\otimes\mathcal{B}(H)\otimes\mathcal{F},\mathcal{B}(H))$*-measurable,* $(\mathcal{F}_t)_{t\geq 0}$*-adapted solution field* $U(t,x,\omega)$ *of the* H*-valued random integral equation*

(1.2.44)

$$U(t,x,\omega)=\phi(t,\omega)(x)+\int_0^t \phi(t-s,\theta(s,\omega))(F(U(s,x,\omega)))\,ds,\quad t\geq 0,\, x\in H,$$

is a mild solution of the semilinear see (1.2.41).

PROOF. Let U be a solution of the random integral equation (1.2.44) with the given measurability properties. It is sufficient to prove that $U(\cdot,x,\cdot)$ satisfies the stochastic integral equation (1.2.42). Substituting from the identity:

$$\phi(t,\omega)(x)=T_t(x)+(\omega)\int_0^t T_{t-s}B\phi(s,\cdot)(x)\,dW(s),\quad t\geq 0, x\in H,$$

into (1.2.44), gives the following a.s. relations

$$\begin{aligned}
U(t,x,\cdot)=&T_t(x)+\int_0^t T_{t-s}B\phi(s,\cdot)(x)\,dW(s)+\int_0^t T_{t-s}(F(U(s,x,\cdot)))\,ds\\
&+\int_0^t\int_0^{t-s} T_{t-s-s'}B\phi(s',\theta(s,\cdot))(F(U(s,x,\cdot)))\,dW(s',\theta(s,\cdot))\,ds\\
=&T_t(x)+\int_0^t T_{t-s}B\phi(s,\cdot)(x)\,dW(s)+\int_0^t T_{t-s}(F(U(s,x,\cdot)))\,ds\\
&+\int_0^t\int_0^{t-s} T_{t-s-s'}B\phi(s',\theta(s,\cdot))(F(U(s,x,\cdot)))\,dW(s'+s)\,ds\\
=&T_t(x)+\int_0^t T_{t-s}B\phi(s,\cdot)(x)\,dW(s)+\int_0^t T_{t-s}(F(U(s,x,\cdot)))\,ds\\
&+\int_0^t\int_s^{t} T_{t-\lambda}B\phi(\lambda-s,\theta(s,\cdot))(F(U(s,x,\cdot)))\,dW(\lambda)\,ds\\
=&T_t(x)+\int_0^t T_{t-s}B\phi(s,\cdot)(x)\,dW(s)+\int_0^t T_{t-s}(F(U(s,x,\cdot)))\,ds\\
&+\int_0^t\int_0^{\lambda} T_{t-\lambda}B\phi(\lambda-s,\theta(s,\cdot))(F(U(s,x,\cdot)))\,ds\,dW(\lambda)\\
=&T_t(x)+\int_0^t T_{t-s}(F(U(s,x,\cdot)))\,ds\\
&+\int_0^t T_{t-\lambda}B\{\phi(\lambda)(x)+\int_0^\lambda \phi(\lambda-s,\theta(s,\cdot))(F(U(s,x,\cdot)))\,ds\}\,dW(\lambda)\\
=&T_t(x)+\int_0^t T_{t-s}(F(U(s,x,\cdot)))\,ds+\int_0^t T_{t-\lambda}BU(\lambda,x,\cdot)\,dW(\lambda)
\end{aligned}$$

for $t\geq 0$. Hence U satisfies (1.2.42) and is therefore a mild solution of the see (1.2.41). □

Our next theorem shows that the random integral equation (1.2.44) admits a unique $(\mathcal{B}(\mathbf{R}^+) \otimes \mathcal{B}(H) \otimes \mathcal{F}, \mathcal{B}(H))$-measurable, $(\mathcal{F}_t)_{t\geq 0}$-adapted solution $U : \mathbf{R}^+ \times H \times \Omega \to H$. The fact that (U, θ) is a smooth perfect cocycle can be read off from (1.2.44), as in the proof of Theorem 1.2.6 below.

For any positive integer j, denote by $L_2^{(j)}(H, H) \subset L^{(j)}(H, H)$ the space of all Hilbert-Schmidt j-multilinear maps $A \in L^{(j)}(H, H)$ given the Hilbert-Schmidt norm

$$\|A\|_{L_2^{(j)}(H,H)} := \sum_{\substack{n_i \geq 1 \\ 1 \leq i \leq j}} |A(e_{n_1}, e_{n_2}, \cdots, e_{n_j})|_H^2 < \infty$$

where $\{e_{n_i} : n_i \geq 1\}$ is a complete orthonormal system in H for each $1 \leq i \leq j$.

THEOREM 1.2.6. *Assume that the operators A, B in* (1.2.41) *satisfy the hypotheses of* Theorem 1.2.4. *Suppose that F satisfies the linear growth and Lipschitz conditions* (1.2.40). *Then the mild solution of* (1.2.41) *has a $(\mathcal{B}(\mathbf{R}^+) \otimes \mathcal{B}(H) \otimes \mathcal{F}, \mathcal{B}(H))$-measurable version $U : \mathbf{R}^+ \times H \times \Omega \to H$ with the following properties:*
(i) *For each $x \in H$, $U(\cdot, x, \cdot) : \mathbf{R}^+ \times \Omega \to H$ is $(\mathcal{F}_t)_{t\geq 0}$-adapted and satisfies* (1.2.42) *a.s.*
(ii) *(U, θ) is a perfect $C^{0,1}$ cocycle (in the sense of* Definition 1.1.2*).*
(iii) *For each $(t, \omega) \in (0, \infty) \times \Omega$, the map $H \ni x \mapsto U(t, x, \omega) \in H$ takes bounded sets into relatively compact sets.*

Moreover, if we assume that F is $C^{k,\epsilon}$ on H for a positive integer k and $\epsilon \in (0, 1]$, then the mild solution (U, θ) also enjoys the following properties:
(iv) *(U, θ) is a $C^{k,\epsilon}$ perfect cocycle.*
(v) *For each $(t, x, \omega) \in \mathbf{R}^+ \times H \times \Omega$, the Fréchet derivatives $D^{(j)}U(t, x, \omega) \in L_2^{(j)}(H, H), 1 \leq j \leq k$, and each map*

$$[0, \infty) \times H \times \Omega \ni (t, x, \omega) \mapsto D^{(j)}U(t, x, \omega) \in L^{(j)}(H, H), \quad 1 \leq j \leq k,$$

is strongly measurable.
(vi) *For any positive a, ρ,*

$$E \log^+ \left\{ \sup_{\substack{0 \leq t_1, t_2 \leq a \\ x \in H}} \frac{|U(t_2, x, \theta(t_1, \cdot))|}{(1 + |x|)} \right\} < \infty$$

and

$$E \log^+ \sup_{\substack{0 \leq t_1, t_2 \leq a \\ |x| \leq \rho, \, 1 \leq j \leq k}} \left\{ \|D^{(j)}U(t_2, x, \theta(t_1, \cdot))\|_{L^{(j)}(H,H)} \right\} < \infty.$$

PROOF. In view of Theorem 1.2.5, we construct a version of the mild solution of (1.2.41) by applying the classical technique of successive approximations to the integral equation (1.2.44). Define the sequence $U_n : \mathbf{R}^+ \times H \times \Omega \to H, n \geq 1$, by

$$\left.\begin{aligned} U_{n+1}(t, x, \omega) &= \phi(t, \omega)(x) + \int_0^t \phi(t - s, \theta(s, \omega))(F(U_n(s, x, \omega)))\, ds, \\ U_1(t, x, \omega) &:= \phi(t, \omega)(x) \end{aligned}\right\} \tag{1.2.45}$$

for all $(t, x, \omega) \in \mathbf{R}^+ \times H \times \Omega$. Fix an arbitrary bounded open set S in H. Let $C_b^0(S, H)$ denote the space of all continuous maps $f : S \to H$ such that $f(S)$ is relatively compact in H. Give $C_b^0(S, H)$ the supremum norm

$$\|f\|_{C_b^0} := \sup_{x \in S} |f(x)|_H, \quad f \in C_b^0(S, H).$$

It is not hard to see that $C_b^0(S,H)$ is a Banach space. For fixed $\omega \in \Omega$ and any $a > 0$, we will view the sequence (1.2.45) as a uniformly convergent sequence of bounded measurable paths $[0,a] \ni t \mapsto U_n(t,\cdot,\omega) \in C_b^0(S,H)$ in the Banach space $C_b^0(S,H)$. To see this, we use induction on n. In view of Theorem 1.2.4 (ii), define the finite random constant $\|\phi\|_\infty := \sup_{0\le s\le t\le a} \|\phi(t-s,\theta(s,\omega))\|_{L(H)}, \omega \in \Omega$. Let C be the positive constant appearing in (1.2.40). Define

$$M_1 := \sup_{x\in S}[|x| + Ca]\|\phi\|_\infty e^{C\|\phi\|_\infty a}, \quad \omega \in \Omega.$$

For integers $n \ge 1$, consider the following induction hypothesis:

Hypotheses H(n):

(i) For each $(t,\omega) \in (0,a] \times \Omega$, $U_n(t,\cdot,\omega) \in C_b^0(S,H)$;
(ii) $|U_n(t,x,\omega)| \le [|x| + Ca]\|\phi\|_\infty e^{C\|\phi\|_\infty t}$ for all $(t,x,\omega) \in [0,a] \times H \times \Omega$;
(iii) $|U_{n+1}(t,x,\omega) - U_n(t,x,\omega)| \le C[1 + \|\phi\|_\infty |x|]L^{n-1}\|\phi\|_\infty^n \frac{t^n}{n!}$, $(t,x,\omega) \in [0,a] \times H \times \Omega$, where L is the Lipschitz constant of F on the ball $B(0,M_1) \subset H$;
(iv) $U_n : \mathbf{R}^+ \times H \times \Omega \to H$ is $(\mathcal{B}(\mathbf{R}^+) \otimes \mathcal{B}(H) \otimes \mathcal{F}, \mathcal{B}(H))$-measurable, and for each $x \in H$, $U_n(\cdot,x,\cdot) : \mathbf{R}^+ \times \Omega \to H$ is $(\mathcal{F}_t)_{t\ge 0}$-adapted.

We will first check that $H(1)$ is satisfied. Since $\phi(t,\cdot,\omega) : H \to H$ is continuous linear for each $(t,\omega) \in [0,a] \times \Omega$, it is clear that $H(1)$(i) and $H(1)$(ii) are satisfied. Using (1.2.45) and the linear growth property of F, it follows that

$$\begin{aligned} |U_2(t,x,\omega) - U_1(t,x,\omega)| &\le C\|\phi\|_\infty \int_0^t [1 + |\phi(s,\omega)(x)|_H]\, ds \\ &\le C[1 + \|\phi\|_\infty |x|]\|\phi\|_\infty t, \end{aligned}$$

for all $(t,x,\omega) \in [0,a] \times H \times \Omega$. Therefore, $H(1)$(iii) holds. To see the measurability (inductive hypothesis $H(1)$(iv)), use the definition of U_1 in (1.2.45) and Theorem 1.2.1.

Now assume that $H(n)$ holds for some integer $n \ge 1$. In particular, for each $(t,\omega) \in (0,a] \times \Omega$, $U_n(t,\cdot,\omega)$ maps S into a relatively compact set in H. Therefore, the map

$$H \ni x \mapsto \int_0^t \phi(t-s,\theta(s,\omega))(F(U_n(s,x,\omega))\, ds \in H$$

takes S into a relatively compact set in H, because, for fixed $(t,\omega) \in (0,a] \times \Omega$, the integrand

$$H \ni x \mapsto \phi(t-s,\theta(s,\omega))(F(U_n(s,x,\omega))) \in H$$

has the same property, and is uniformly bounded in $(s,x) \in [0,t] \times S$ ($H(n)$(ii)). Hence, $U_{n+1}(t,\cdot,\omega)(S)$ is relatively compact in H for each $(t,\omega) \in (0,a] \times \Omega$. Since $U_n(t,\cdot,\omega) : H \to H$ is continuous, it is easy to see from (1.2.45) that $U_{n+1}(t,\cdot,\omega) : H \to H$ is also continuous for each $(t,\omega) \in (0,a] \times \Omega$. Hence, $H(n+1)$(i) is satisfied. Using $H(n)$(ii), the Lipschitz property of F and (1.2.45), a straightforward computation shows that $H(n+1)$(iii) is satisfied. A similar argument, using $H(n)$(ii), the linear growth property of F and (1.2.45), shows that $H(n+1)$(ii) also holds. To check $H(n+1)$(iv), note first that for fixed $s \in [0,t]$, the map $\Omega \ni \omega \mapsto \phi(t-s,\theta(s,\omega)) \in L(H)$ is $\mathcal{F}_t$-measurable. This follows from the approximation argument at the end of the proof of Theorem 1.2.4. Hence by $H(n)(iv)$, it follows that for fixed $s \in [0,t]$, the map $\Omega \ni \omega \mapsto \phi(t-s,\theta(s,\omega))(F(U_n(s,x,\omega))) \in$

$L(H)$ is $\mathcal{F}_t$-measurable. Hence by (1.2.45), it is easy to see that $U_{n+1}(t,x,\cdot)$ is $\mathcal{F}_t$-measurable for fixed $(t,x) \in \mathbf{R}^+ \times H$. Furthermore, the integrand on the right-hand-side of (1.2.45) is jointly-measurable in (s,x,ω), and therefore $U_{n+1}(t,\cdot,\cdot)$ is jointly measurable for any fixed $t > 0$. By continuity of the path $\mathbf{R}^+ \ni t \mapsto U_{n+1}(t,x,\omega)$ for fixed $(x,\omega) \in H \times \Omega$, the joint measurability of $U_{n+1} : \mathbf{R}^+ \times H \times \Omega \to H$ follows. Hence $H(n+1)$(iv) is satisfied. Therefore, $H(n)$ holds by induction for all integers $n \geq 1$.

The inequality $H(n)$(iii) implies that the series $\sum_{n=1}^{\infty} [U_{n+1}(t,\cdot,\omega) - U_n(t,\cdot,\omega)]$ converges in $C_b^0(S,H)$ uniformly in $t \in [0,a]$ for each $\omega \in \Omega$. Therefore, the sequence $\{U_n(t,\cdot,\omega)\}_{n=1}^{\infty}$ converges in $C_b^0(S,H)$ uniformly in $t \in [0,a]$ for each $\omega \in \Omega$. Its limit

$$\lim_{n\to\infty} U_n(t,\cdot,\omega) = U_1(t,\cdot,\omega) + \sum_{n=1}^{\infty} [U_{n+1}(t,\cdot,\omega) - U_n(t,\cdot,\omega)], \quad (t,\omega) \in [0,a] \times \Omega,$$

is a solution of the random integral equation (1.2.44). Call this limit $U(t,\cdot,\omega) \in C_b^0(S,H)$ for $(t,\omega) \in \mathbf{R}^+ \times \Omega$. It is immediately clear from $H(n)$(iv) and Theorem 1.2.5 that U satisfies the measurability requirements and assertion (i) of the theorem.

We next show that $U(t,\cdot,\omega) : H \to H$ is C^1 for fixed $(t,\omega) \in \mathbf{R}^+ \times \Omega$. For each $(x,y,\omega) \in H \times H \times \Omega$, denote by $z(\cdot,x,y,\omega)$ the unique solution of the random linear integral equation:

$$\begin{aligned} z(t,x,y,\omega) &= \int_0^t \phi(t-s,\theta(s,\omega))DF(U(s,x,\omega))z(s,x,y,\omega)\,ds \\ &\quad + \int_0^t \phi(t-s,\theta(s,\omega))DF(U(s,x,\omega))\phi(s,\omega)(y)\,ds, \quad t > 0. \end{aligned} \tag{1.2.46}$$

If we suppress $y \in H$, we can view (1.2.46) as a linear integral equation in $L_2(H)$ with a unique solution $[0,\infty) \ni t \mapsto z(t,x,\cdot,\omega) \in L_2(H)$ for fixed $(x,\omega) \in H \times \Omega$. This holds easily (by successive approximations) because DF is bounded on bounded subsets of H and $\{U(t,x,\omega); 0 \leq t \leq a, |x| \leq M\}$ is bounded for any $M > 0$, and $\|\phi\|_\infty$ is finite. We claim that $U(t,\cdot,\omega)$ is Fréchet differentiable with Fréchet derivative $DU(t,x,\omega) \in L_2(H)$ given by

$$DU(t,x,\omega)(y) = z(t,x,y,\omega) + \phi(t,\omega)(y), \quad y \in H \tag{1.2.47}$$

for each $(t,x,\omega) \in \mathbf{R}^+ \times H \times \Omega$. To prove our claim, define

$$\mu(t,x,y,h,\omega) := U(t,x+hy,\omega) - U(t,x,\omega) - h[z(t,x,y,\omega) + \phi(t,\omega)(y)] \tag{1.2.48}$$

for each $(t,x,y,h,\omega) \in \mathbf{R}^+ \times H \times H \times \mathbf{R} \times \Omega$. Using (1.2.48), (1.2.44) and (1.2.46), we obtain:
(1.2.49)

$$\begin{aligned}\mu(t,x,y,h,\omega) = \int_0^t & \phi(t-s,\theta(s,\omega))DF(U(s,x,\omega))\mu(s,x,y,h,\omega)\,ds \\ &+ \int_0^t \phi(t-s,\theta(s,\omega))\Big\{\int_0^1 DF[\lambda U(s,x+hy,\omega) \\ &\qquad + (1-\lambda)U(s,x,\omega)] - DF(U(s,x,\omega))\,d\lambda\Big\} \\ &\qquad \cdot (U(s,x+hy,\omega) - U(s,x,\omega))\,ds\end{aligned}$$

for all $(t,x,y,h,\omega) \in \mathbf{R}^+ \times H \times H \times \mathbf{R} \times \Omega$. Set

$$M_2 := \sup_{\substack{|h|\le 1, |y|\le 1 \\ 0\le s\le a}} \{|U(s,x+hy,\omega)|\}, \quad \omega \in \Omega.$$

Then M_2 is finite for each $\omega \in \Omega$, because of $H(n)$(ii). Let $L_1 > 0$ be the Lipschitz constant of DF on the ball $B(0,M_2)$, and $\|DF\|$ be the bound of DF on $B(0,M_2)$. Then (1.2.49) implies the following inequality:

$$\begin{aligned}|\mu(t,x,y,h,\omega)| \le \|\phi\|_\infty \|DF\| & \int_0^t |\mu(s,x,y,\omega)|\,ds \\ &+ L_1\|\phi\|_\infty \int_0^t |U(s,x+hy,\omega) - U(s,x,\omega)|^2\,ds \end{aligned} \tag{1.2.50}$$

for all $t \in [0,a], x,y \in H, h \in \mathbf{R}, |y|, |h| \le 1, \omega \in \Omega$. Using (1.2.44) and Gronwall's lemma, it is easy to see that

$$|U(t,x+hy,\omega) - U(t,x,\omega)| \le |h|\|\phi\|_\infty |y| e^{\|\phi\|_\infty \|DF\| t} \tag{1.2.51}$$

for all $t \in [0,a], x,y \in H, h \in \mathbf{R}, |y|, |h| \le 1, \omega \in \Omega$. By (1.2.50), (1.2.51) and another simple application of Gronwall's lemma, we obtain

$$|\mu(t,x,y,h,\omega)| \le \frac{|h|^2|y|^2\|\phi\|_\infty^2 L_1}{2\|DF\|}\Big[e^{2\|\phi\|_\infty\|DF\|a} - 1\Big] e^{\|\phi\|_\infty\|DF\|t} \tag{1.2.52}$$

for all $t \in [0,a], x,y \in H, h \in \mathbf{R}, |y|, |h| \le 1, \omega \in \Omega$. Thus,

$$\lim_{h\to 0} \frac{1}{h} \sup_{\substack{|y|\le 1 \\ 0\le t\le a}} |\mu(t,x,y,h,\omega)| = 0 \tag{1.2.53}$$

for all $x \in H, \omega \in \Omega$. The above relation shows that $U(t,\cdot,\omega) : H \to H$ is Fréchet differentiable at any $x \in H$ and our claim (1.2.47) holds. Now combining (1.2.46) and (1.2.47), it follows that $DU(t,x,\omega)$ satisfies the $L(H)$-valued integral equation:

$$DU(t,x,\omega) = \phi(t,\omega) + \int_0^t \phi(t-s,\theta(s,\omega))DF(U(s,x,\omega))DU(s,x,\omega)\,ds \tag{1.2.54}$$

for each $(t,x,\omega) \in \mathbf{R}^+ \times H \times \Omega$. In the above integral equation, the "coefficients"

$$[0,\infty) \times \Omega \ni (t,\omega) \mapsto \phi(t,\omega) \in L(H)$$
$$\Delta \times H \times \Omega \ni (s,t,x,\omega) \mapsto \phi(t-s,\theta(s,\omega))DF(U(s,x,\omega)) \in L(H)$$

are jointly measurable, where $\Delta = \{(s,t) \in \mathbf{R}^2 : 0 \le s \le t\}$. Therefore, the solution map

$$[0,\infty) \times H \times \Omega \ni (t,x,\omega) \mapsto DU(t,x,\omega) \in L(H)$$

is jointly measurable. Furthermore, by continuity of the map

$$H \ni x \mapsto DF(U(s,x,\omega)) \in L(H,\mathbf{R})$$

it follows from (1.2.54) that the map $H \ni x \mapsto DU(t,x,\omega) \in L(H)$ is continuous for fixed $t > 0$ and $\omega \in \Omega$. Thus $U(t,\cdot,\omega) : H \to H$ is C^1. (In fact, the map $H \ni x \mapsto DU(t,x,\omega) \in L_2(H), t > 0$, is continuous because of the continuity of the map $H \ni x \mapsto z(t,x,\cdot,\omega) \in L_2(H)$ in the $L_2(H)$-valued integral equation underlying (1.2.46).)

Suppose further that F is $C^{k,\epsilon}, k \ge 1, \epsilon \in (0,1]$. For $k = 1$, assertion (vi) of the theorem follows from (1.2.44), the linear growth property of F, (1.2.54), Gronwall's lemma and the fact that $E\|\phi\|_\infty < \infty$. By suppressing y in (1.2.46) and taking higher-order Fréchet derivatives with respect to x of the underlying $L_2(H)$-valued integral equation, assertions (v) and (vi) can be established by induction on $k > 1$.

It remains to prove that (U,θ) is a perfect cocycle on H. We use uniqueness of solutions of (1.2.44). Fix $t_1 \ge 0$, $\omega \in \Omega$ and $x \in H$. It is sufficient to prove that

$$U(t+t_1,x,\omega) = U(t,U(t_1,x,\omega),\theta(t_1,\omega)) \tag{1.2.55}$$

for all $t \ge 0$. Define the two mappings $y, z : [0,\infty) \to H$ by

$$y(t) := U(t,U(t_1,x,\omega),\theta(t_1,\omega)), \quad z(t) := U(t+t_1,x,\omega) \tag{1.2.56}$$

for all $t \ge 0$. Since U satisfies (1.2.44), it follows that

$$\begin{aligned} y(t) =& \phi(t,\theta(t_1,\omega))(U(t_1,x,\omega)) \\ &+ \int_0^t \phi(t-s,\theta(s,\theta(t_1,\omega)))(F(U(s,U(t_1,x,\omega),\theta(t_1,\omega)))\,ds \\ =& \phi(t,\theta(t_1,\omega))(\phi(t_1,\omega)(x)) \\ &+ \int_0^{t_1} \phi(t,\theta(t_1,\omega))\{\phi(t_1-s,\theta(s,\omega))(F(U(s,x,\omega)))\}\,ds \\ &+ \int_{t_1}^{t+t_1} \phi(t+t_1-s,\theta(s,\omega))(F(y(s-t_1)))\,ds \\ =& \phi(t+t_1,\omega)(x) + \int_0^{t_1} \phi(t+t_1-s,\theta(s,\omega))(F(U(s,x,\omega)))\,ds \\ &+ \int_{t_1}^{t+t_1} \phi(t+t_1-s,\theta(s,\omega))(F(y(s-t_1)))\,ds \end{aligned}$$

for all $t \ge 0$. Making the substitution $t' := t + t_1$, the above relation yields

$$\begin{aligned} y(t'-t_1) =& \phi(t',\omega)(x) + \int_0^{t_1} \phi(t'-s,\theta(s,\omega))(F(U(s,x,\omega)))\,ds \\ &+ \int_{t_1}^{t'} \phi(t'-s,\theta(s,\omega))(F(y(s-t_1)))\,ds, \quad t' > t_1. \end{aligned} \tag{1.2.57}$$

Using (1.2.44) and the definition of z, it follows that

$$z(t) = \phi(t+t_1,\omega)(x) + \int_0^{t_1} \phi(t+t_1-s,\theta(s,\omega))(F(U(s,x,\omega)))\,ds$$
$$+ \int_{t_1}^{t+t_1} \phi(t+t_1-s,\theta(s,\omega))(F(U(s,x,\omega)))\,ds \quad t \geq 0.$$

Therefore,

$$z(t'-t_1) = \phi(t',\omega)(x) + \int_0^{t_1} \phi(t'-s,\theta(s,\omega))(F(U(s,x,\omega)))\,ds$$
$$+ \int_{t_1}^{t'} \phi(t'-s,\theta(s,\omega))(F(z(s-t_1)))\,ds, \quad t' \geq t_1. \tag{1.2.58}$$

It is easy to see that (1.2.57) and (1.2.58) imply

$$|y(t'-t_1) - z(t'-t_1)| \leq \int_{t_1}^{t'} \|\phi(t'-s,\theta(s,\omega))\| \cdot |F(y(s-t_1)) - F(z(s-t_1))|\,ds$$
$$\leq L\|\phi\|_\infty \int_{t_1}^{t'} |y(s-t_1)) - z(s-t_1)|\,ds, \quad t_1 \leq t' \leq t_1 + a, \tag{1.2.59}$$

where L is the Lipschitz constant of F on the bounded set $\{y(s), z(s), 0 \leq s \leq a\}$. From the above inequality, we get $y(t'-t_1) - z(t'-t_1) = 0$ for all $t' \geq t_1$. Hence, $y(t) = z(t)$ for all $t \geq 0$. This implies the perfect cocycle property (1.2.55) and completes the proof of the theorem. □

REMARKS.

(i) From the proof of Theorem 1.2.6, it is easy to see that the assertions of the theorem still hold if one replaces the linear growth condition on F by the condition that F carries bounded sets in H into bounded sets, and $U(\cdot,\cdot,\omega)$ is bounded on bounded subsets of $[0,\infty) \times H$.

(ii) In (1.2.41), it is possible to replace F by a time-dependent $\tilde{F} : \mathbf{R}^+ \times H \to H$ of class $C^{k,\epsilon}$ in the second variable uniformly with respect to t in compacta. This gives a $C^{k,\epsilon}$ semiflow $V : \Delta \times H \times \Omega \to H$ in the sense of Definition 1.1.1.

EXAMPLE. Let $\mathcal{D}$ denote the bounded domain in $\mathbf{R}^d$ defined by

$$\mathcal{D} := \{(x_1, x_2, \cdots, x_d); 0 \leq x_i \leq \pi, 1 \leq i \leq d\}.$$

Let Δ be the Laplacian on $\mathcal{D}$. Equip the operator $B := -\Delta + I$ on $\mathcal{D}$ with a Neumann boundary condition, and consider the following spde:

$$du(t,x) = -Au(t,x)\,dt + f(u(t,x))\,dt + c(x)u(t,x)\,dW(t) \tag{1.2.60}$$

where W stands for standard (one-dimensional) Brownian motion, and $A := B^\alpha$ for some positive constant α. Let $\phi_0(x), \phi_n(x)$ be functions on $[0,\pi]$ defined by

$$\phi_0(x) = \left(\frac{1}{\pi}\right)^{\frac{1}{2}}, \qquad \phi_n(x) = \left(\frac{2}{\pi}\right)^{\frac{1}{2}} \cos(nx), \quad n \geq 1.$$

For any non-negative integers $i_1, i_2, ..., i_d$, define

$$\psi_{i_1,i_2,\ldots,i_d}(x_1, x_2, .., x_d) := \phi_{i_1}(x_1)\phi_{i_2}(x_2)\cdots\phi_{i_d}(x_d) \tag{1.2.61}$$

for $x_i \in [0,\pi], 1 \le i \le d$. Then the family $\{\psi_{i_1,i_2,...,i_d} : 0 \le i_1, i_2, ..., i_d < \infty\}$ forms an orthonormal basis of $L^2(\mathcal{D})$. It is easy to verify that each $\psi_{i_1,i_2,...,i_d}$ is an eigenfunction of B with corresponding eigenvalue

$$\lambda_{i_1,i_2,...,i_d} = 1 + \sum_{j=1}^{d} i_j^2.$$

Thus

$$A\psi_{i_1,i_2,...,i_d} = \lambda^{\alpha}_{i_1,i_2,...,i_d}\psi_{i_1,i_2,...,i_d}$$

If $\alpha > \frac{d}{2}$, we have

$$\begin{aligned}\sum_{i_1,i_2,\cdots,i_d=0}^{\infty} \frac{1}{\lambda^{\alpha}_{i_1,i_2,...,i_d}} &= \sum_{i_1,i_2,\cdots,i_d=0}^{\infty} \left(1+\sum_{j=1}^{d} i_j^2\right)^{-\alpha} \\ &\le \sum_{i_1,i_2,\cdots,i_d=0}^{\infty} \prod_{j=1}^{d}\left(\frac{1}{d}+i_j^2\right)^{-\frac{\alpha}{d}} \\ &= \left\{\sum_{k=0}^{\infty}\left(\frac{1}{d}+k^2\right)^{-\frac{\alpha}{d}}\right\}^d < \infty.\end{aligned}$$

If f is Lipschitz and $c(x)$ is bounded, then the assumptions in Theorems 1.2.1 and 1.2.6 are satisfied. So Theorem 1.2.6 applies, and the spde (1.2.60) admits a $C^{0,1}$ perfect cocycle $U : \mathbf{R}^+ \times L^2(\mathcal{D}) \times \Omega \to L^2(\mathcal{D})$ satisfying assertions (i)-(iii) of Theorem 1.2.6.

In the next section, we will see more applications of the results established in this section.

1.3. Semilinear spde's: Lipschitz nonlinearity

Let $\mathcal{D}$ be a smooth bounded domain in $\mathbf{R}^d$. Consider the Laplacian operator:

$$\Delta := \sum_{i=1}^{d} \frac{\partial^2}{\partial \xi_i^2} \tag{1.3.1}$$

defined on $\mathcal{D}$. Let $L^\infty(\mathcal{D})$ stand for all essentially bounded measurable functions $\psi : \mathcal{D} \to \mathbf{R}$ with the usual norm

$$\|\psi\|_\infty := \mathrm{essup}_{\xi\in\mathcal{D}}|\psi(\xi)|.$$

Denote by $C_0^\infty(\mathcal{D})$ the set of all smooth test functions $\phi : \mathcal{D} \to \mathbf{R}$ which vanish on $\partial\mathcal{D}$. Let $H := H_0^k(\mathcal{D})$ be the Sobolev space of order $k > d/2$, i.e., the completion of $C_0^\infty(\mathcal{D})$ under the Sobolev norm

$$||u||^2_{H_0^k} := \sum_{|\alpha|\le k} \int_{\mathcal{D}} |D^\alpha u(\xi)|^2 d\xi,$$

where $d\xi$ denotes d-dimensional Lebesgue measure on $\mathbf{R}^d$.

Consider the spde

$$
\left.\begin{aligned}
du(t) =& \frac{1}{2}\Delta u(t)dt + f(u(t))dt + \sum_{i=1}^{\infty} \sigma_i u(t)\, dW^i(t), \quad t > 0\\
u(0) =& \psi \in H_0^k(\mathcal{D})\\
u(t)|_{\partial\mathcal{D}} =& 0, \quad t \geq 0,
\end{aligned}\right\} \tag{1.3.2}
$$

where $f : \mathbf{R} \to \mathbf{R}$ is a C_b^∞ function, $\sigma_i : \mathcal{D} \to \mathbf{R}, i \geq 1$, are functions in the Sobolev space $H_0^s(\mathcal{D})$ with $s > k + \frac{d}{2}$, and $W^i, i \geq 1$, are standard independent one-dimensional Brownian motions on a complete filtered Wiener space $(\Omega, \bar{\mathcal{F}}, (\mathcal{F})_{t\geq 0}, P)$. Assume that the coefficients σ_i in (1.3.2) satisfy the following condition

$$
\sum_{i=1}^{\infty} \|\sigma_i\|_{H_0^s}^2 < \infty. \tag{1.3.3}
$$

An $(\mathcal{F}_t)_{t\geq 0}$-adapted random field $u : \mathbf{R}^+ \times \mathcal{D} \times \Omega \to \mathbf{R}$ is a *weak solution* of (1.3.2) if $u(t, \cdot, \omega) \in H_0^k(\mathcal{D})$ for a.a. $\omega \in \Omega, t \geq 0$, and the following identity holds:

$$
\left.\begin{aligned}
d < u(t), \phi >_{L^2} =& \nu < u(t), \Delta\phi >_{L^2} dt + < f(u(t)), \phi >_{L^2} dt\\
& + \sum_{i=1}^{\infty} < \sigma_i u(t), \phi >_{L^2} dW^i(t),\\
u(0) =& \psi \in H_0^k(\mathcal{D}),\\
u(t)|_{\partial\mathcal{D}} =& 0,
\end{aligned}\right\}
$$

for all $\phi \in C_0^\infty(\mathcal{D})$ a.s. and all $t > 0$. In the above equality, $< \cdot, \cdot >_{L^2}$ denotes the inner product on the Hilbert space $L^2(\mathcal{D})$ of all square-integrable functions $\psi : \mathcal{D} \to \mathbf{R}$, viz.

$$
< \psi_1, \psi_2 >_{L^2} := \int_{\mathcal{D}} \psi_1(\xi)\psi_2(\xi)\, d\xi, \quad \psi_1, \psi_2 \in L^2(\mathcal{D}).
$$

Recall that $d\xi$ stands for d-dimensional Lebesgue measure.

We will show that (1.3.2) admits a unique weak solution $u(t) \in H = H_0^k(\mathcal{D})$ a.s., $t > 0$, for each $\psi \in H$. Furthermore, the ensemble of all weak solutions of (1.3.2) generates a C^∞ perfect cocycle (*also denoted by the same symbol*) $u : \mathbf{R}^+ \times H \times \Omega \to H$, satisfying the assertions of Theorem 1.3.5 below. In particular, the stochastic semiflow $u(t, \cdot, \omega) : H \to H$ takes bounded sets into relatively compact sets in H.

In this section and for the rest of the article, we should emphasize that although the weak *solution* $u : \mathbf{R}^+ \times \mathcal{D} \times \Omega \to \mathbf{R}$ of (1.3.2) and the associated stochastic *semiflow* $u : \mathbf{R}^+ \times H \times \Omega \to H$ are denoted by the same symbol u, the distinction between the two notions should be clear from the context.

Set $A := -\frac{1}{2}\Delta$ with Dirichlet boundary conditions on $\partial\mathcal{D}$. We will view the spde (1.3.2) as a semilinear see in H of the form (1.2.41) (Section 1.2). First, define the Nemytskii operator

$$
F(u)(\xi) := f(u(\xi)), \quad \xi \in \mathcal{D}, u \in H. \tag{1.3.3'}
$$

In Lemma 1.3.3 below, we will show that F is a C^∞ map $H \to H$. Secondly, apply the Gramm-Schmidt orthogonalization process to the sequence $\{\sigma_i\}_{i=1}^\infty$ in $H_0^s(\mathcal{D})$. This gives an orthonormal family $\{f_i\}_{i=1}^\infty$ in $H_0^s(\mathcal{D})$. Denote by K the closed linear span of $\{f_i\}_{i=1}^{\infty}$ in $H_0^s(\mathcal{D})$. The reader may check that K is a closed subspace of the Hilbert space $H_0^s(\mathcal{D})$, and there is a separable Hilbert space E such that $K \subset E$ is a Hilbert-Schmidt embedding (e.g. take $E = L^2(\mathcal{D})$).

Define the process

$$W(t) := \sum_{i=1}^{\infty} W^i(t) f_i, \quad t > 0.$$

Then it follows from (1.3.3) that W is an E-valued cylindrical Brownian motion on the canonical complete filtered Wiener space $(\Omega, \bar{\mathcal{F}}, (\mathcal{F}_t)_{t\in\mathbf{R}^+}, P)$, and with covariance space K (cf. section 1.2). Denote by $\theta : \mathbf{R}^+ \times \Omega \to \Omega$ the standard P-preserving (ergodic) Brownian shift. It is easy to see that (W, θ) is a perfect helix on E:

$$W(t_1 + t_2, \omega) = W(t_2, \theta(t_1, \omega)) - W(t_1, \omega), \quad t_1, t_2 \in \mathbf{R}^+, \omega \in \Omega.$$

Define the continuous linear operator $B : H \to L_2(K, H)$ by setting

$$B(u)(f_i) := \sigma_i u, \quad u \in H = H_0^k(\mathcal{D}),\ i \geq 1.$$

In view of the continuous linear (Sobolev) embedding

$$H_0^s(\mathcal{D}) \hookrightarrow C^k(\mathcal{D}),$$

it is easy to see that $B \in L(H, L_2(K, H))$ and satisfies Condition (B) of section 1.2(a). Thirdly, observe that weak solutions of the spde (1.3.2) correspond to mild solutions of the semilinear see:

$$\left.\begin{aligned} du(t) =& -Au(t)dt + F(u(t))dt + Bu(t)\,dW(t), \quad t > 0 \\ u(0) =& \psi \in H := H_0^k(\mathcal{D}) \end{aligned}\right\} \tag{1.3.2$'$}$$

([D-Z.1], p. 156).

Finally, we will establish a perfect C^∞-cocycle on the Sobolev space $H = H_0^k(\mathcal{D})$ for mild solutions of the semilinear see (1.3.2$'$), and hence for weak solutions of the spde (1.3.2).

We begin with some preparation. Following standard notation, let α be a d-tuple of non-negative integers, viz. $\alpha := (\alpha_1, \alpha_2, \cdots, \alpha_d)$ and denote $|\alpha| := \alpha_1 + \alpha_2 + \cdots + \alpha_d$. For any $\phi \in C^{|\alpha|}(\mathcal{D})$, denote

$$(D^{(\alpha)}\phi)(\xi) \equiv \phi^{(\alpha)}(\xi) := \partial_1^{\alpha_1}\partial_2^{\alpha_2}\cdots\partial_d^{\alpha_d}\phi(\xi), \quad \xi \in \mathcal{D},$$

and for any integer $l > 0$, define

$$\|D^l\phi\|_{L^2} := \sum_{|\alpha|=l} \|D^{(\alpha)}\phi\|_{L^2}.$$

LEMMA 1.3.1. *Let $\beta_1, \cdots, \beta_\mu$ be d-tuples and $|\alpha| = |\beta_1| + |\beta_2| + \cdots + |\beta_\mu|$, then there exists a constant $c > 0$ such that*

$$\begin{aligned} &\|f_1^{(\beta_1)} f_2^{(\beta_2)} \cdots f_\mu^{(\beta_\mu)}\|_{L^2} \\ &\leq c^\mu \|f_1\|_{L^\infty}^{1-\frac{|\beta_1|}{|\alpha|}} \|f_2\|_{L^\infty}^{1-\frac{|\beta_2|}{|\alpha|}} \cdots \|f_\mu\|_{L^\infty}^{1-\frac{|\beta_\mu|}{|\alpha|}} \|D^{|\alpha|}f_1\|_{L^2}^{\frac{|\beta_1|}{|\alpha|}} \\ &\qquad \cdot \|D^{|\alpha|}f_2\|_{L^2}^{\frac{|\beta_2|}{|\alpha|}} \|D^{|\alpha|}f_\mu\|_{L^2}^{\frac{|\beta_\mu|}{|\alpha|}}, \end{aligned}$$

where $f_j \in C^{|\alpha|}(\mathcal{D})$, $1 \leq j \leq \mu$.

A proof of this lemma is given in ([Ta], pp. 9-10), using Gagliardo-Nirenberg-Moser estimates.

LEMMA 1.3.2. *In* (1.3.2), *let* f *be* $C^\infty, f(0) = 0$, *and* F *be the Nemytskii operator* (1.3.3′). *Then there is a positive constant* c *such that*

$$\|F(u)\|_{H_0^k(\mathcal{D})} \leq cC_k(\|u\|_{L^\infty})(1+\|u\|_{L^\infty})^{k-1}\|u\|_{H_0^k(\mathcal{D})},$$

for all $u \in H_0^k(\mathcal{D})$, *where*

$$C_k(\lambda) := \sup_{\substack{1\leq\mu\leq k \\ |z|\leq\lambda}} |f^{(\mu)}(z)|.$$

PROOF. We need only prove the assertion of the lemma for $u \in C_0^\infty(\mathcal{D})$. The chain rule gives for any d-tuple α with $1 \leq |\alpha| \leq k$,

$$D^\alpha F(u) = \sum_{\substack{\beta_1+\beta_2+\cdots+\beta_\mu=\alpha \\ 1\leq\mu\leq|\alpha|}} c_\beta u^{(\beta_1)}u^{(\beta_2)}\cdots u^{(\beta_\mu)}(f^{(\mu)}\circ u).$$

Hence

$$\|D^\alpha F(u)\|_{L^2} \leq C_k(\|u\|_{L^\infty}) \cdot \sum_{\substack{\beta_1+\beta_2+\cdots+\beta_\mu=\alpha, \\ 1\leq\mu\leq|\alpha|}} c_\beta\|u^{(\beta_1)}u^{(\beta_2)}\cdots u^{(\beta_\mu)}\|_{L^2}.$$

Applying Lemma 1.3.1 to $f_i = u, i = 1, 2, \cdots, \mu$, we have

$$\|u^{(\beta_1)}u^{(\beta_2)}\cdots u^{(\beta_\mu)}\|_{L^2} \leq c^\mu\|u\|_{L^\infty}^{\mu-1}\|D^{|\alpha|}u\|_{L^2}.$$

Therefore,

$$\begin{aligned}\sum_{\substack{\beta_1+\beta_2+\cdots+\beta_\mu=\alpha, \\ 1\leq\mu\leq|\alpha|}} & c_\beta\|u^{(\beta_1)}u^{(\beta_2)}\cdots u^{(\beta_\mu)}\|_{L^2} \\ &\leq \sum_{\substack{\beta_1+\beta_2+\cdots+\beta_\mu=\alpha, \\ 1\leq\mu\leq|\alpha|}} c_\beta c^\mu\|u\|_{L^\infty}^{\mu-1}\|D^{|\alpha|}u\|_{L^2} \\ &\leq c\|D^{|\alpha|}u\|_{L^2}\sum_{1\leq\mu\leq|\alpha|} C_\mu^{|\alpha|}\|u\|_{L^\infty}^{\mu-1} \\ &\leq c\|D^{|\alpha|}u\|_{L^2}(1+\|u\|_{L^\infty})^{|\alpha|-1} \\ &\leq c\|u\|_{H_0^k}(1+\|u\|_{L^\infty})^{k-1},\end{aligned}$$

for a constant $c > 0$. Note also that

$$\|F(u)\|_{L^2} \leq C_k\|u\|_{L^2} \leq C_k\|u\|_{H_0^k}, \quad u \in C_0^\infty(\mathcal{D}).$$

The assertion of the lemma follows easily from the above inequality. □

LEMMA 1.3.3. *Suppose* $k > \frac{d}{2}$, *and* $f : \mathbf{R} \to \mathbf{R}$ *is a* C^∞ *function. Then the Nemytskii operator* $F : H_0^k(\mathcal{D}) \to H_0^k(\mathcal{D})$ *defined by* (1.3.3′) *is a* C^∞ *map from* $H_0^k(\mathcal{D})$ *into* $H_0^k(\mathcal{D})$.

PROOF. Recall the following Sobolev embeddings

$$H_0^r(\mathcal{D}) \hookrightarrow L^{\frac{2d}{d-2r}}(\mathcal{D}), \qquad r < \frac{d}{2},$$

$$H_0^r(\mathcal{D}) \hookrightarrow L^{\infty}(\mathcal{D}), \qquad r > \frac{d}{2}.$$

Let us first prove that $F \in C^1(H, H)$, where $H := H_0^k(\mathcal{D})$. Fix $u \in H$. We will show that F is Fréchet differentiable and $DF(u)(h)(\xi) \equiv S_u(h)(\xi) = f'(u(\xi))h(\xi)$, $h \in H, \xi \in \mathcal{D}$. To prove this, note that only functions in some ball $B(0, \delta) \subset H$ centered at 0 are involved. By the Sobolev embedding theorem, the range of functions in $B(0, \delta)$ is contained in a compact interval in $\mathbf{R}$. Thus, we can assume $f \in C_b^{\infty}$ in the sequel. We start by proving that $S_u(h) \in H$ for $h \in H$. Let $r \le k$. By the chain and product rules, it follows that $(S_u(h))^{(r)}$ can be written as a finite sum whose general term is of the form: $C(\xi)u^{(l_1)}(\xi) \cdots u^{(l_m)}(\xi)h^{(j_1)}(\xi) \cdots h^{(j_n)}(\xi)$, where $C(\cdot) \in L^{\infty}(\mathcal{D})$, and $l_1 + \cdots + l_m + j_1 + \cdots j_n = r$. Since $u^{(l)} \in H_0^{k-l}(\mathcal{D})$ and $h^{(j)} \in H_0^{k-j}(\mathcal{D})$, the Sobolev embedding theorem implies that $u^{(l)} \in L^{\frac{2d}{d-2k+2l}}(\mathcal{D})$ and $h^{(j)} \in L^{\frac{2d}{d-2k+2j}}(\mathcal{D})$. As

$$\sum_{i=1}^{m}(d - 2k + 2l_i) + \sum_{i=1}^{n}(d - 2k + 2j_i) - d \le (m + n - 1)(d - 2k) < 0,$$

we have

$$\frac{\sum_{i=1}^{m}(d - 2k + 2l_i)}{2d} + \frac{\sum_{i=1}^{n}(d - 2k + 2j_i)}{2d} \le \frac{1}{2}.$$

By Hölder's inequality and the Sobolev embedding theorem, this implies that

$$|C(\cdot)u^{(l_1)}(\cdot) \cdots u^{(l_m)}(\cdot)h^{(j_1)}(\cdot) \cdots h^{(j_n)}(\cdot)|_{L^2(\mathcal{D})} \le c|u|_H^m |h|_H^n.$$

where c is a positive constant. Thus $S_u(h)$ is not only in H, but the map $H \ni h \mapsto S_u(h) \in H$ is a continuous linear operator. Now

$$F(u + th)(\xi) - F(u)(\xi) - tS_u(h)(\xi) = \int_0^t [f'(u(\xi) + sh(\xi)) - f'(u(\xi))]h(\xi)ds$$

for each $\xi \in \mathcal{D}, u, h \in H, t \ge 0$. To show that $DF(u) = S_u$, we need to prove that

$$\lim_{t \to 0} \sup_{|h|_H \le 1} \left| \frac{1}{t} \int_0^t [f' \circ (u + sh) - f' \circ (u)] \cdot h \, ds \right|_H = 0.$$

It is sufficient to establish

$$\lim_{s \to 0} \sup_{|h|_H \le 1} |[f' \circ (u + sh) - f' \circ (u)] \cdot h|_H = 0.$$

The above relation will hold if we show that

$$\lim_{s \to 0} \sup_{|h|_H \le 1} |[(f' \circ (u + sh) - f' \circ (u)) \cdot h]^{(r)}|_{L^2(\mathcal{D})} = 0$$

for $r \le k$.

Elementary computations show that $[(f' \circ (u + sh) - f' \circ (u)) \cdot h]^{(r)}(\xi)$ is a finite sum consisting of terms which are either of the form

$$G_1(\xi, h, s) := (f^{(l)} \circ (u + sh) - f^{(l)} \circ (u))(\xi)u^{(l_1)}(\xi)...u^{(l_m)}(\xi)h^{(j_1)}(\xi)...h^{(j_n)}(\xi)$$

for $l \leq r+1$, or of the form

$$G_2(\xi, h, s) := s^a C(\xi) u^{(l_1)}(\xi) \ldots u^{(l_m)}(\xi) h^{(j_1)}(\xi) \ldots h^{(j_n)}(\xi)$$

where $a \geq 1, C(\cdot) \in L^\infty(\mathcal{D}), l_1 + \ldots + l_m + j_1 + \ldots + j_n = r,\ h \in H,\ s > 0$. For terms like G_1, using the Lipschitz continuity of $f^{(l)}$, it follows that

$$|G_1(\xi, h, s)| \leq cs|h|_{L^\infty(\mathcal{D})} |u^{(l_1)}(\xi) \ldots u^{(l_m)}(\xi) h^{(j_1)}(\xi) \ldots h^{(j_n)}(\xi)|$$

for $\xi \in \mathcal{D}, h \in H, s > 0$. Using Hölder's inequality and the Sobolev embedding theorem, and arguing as in the proof of $S_u(h) \in H$, we obtain the following estimate

$$\|G_1(\cdot, h, s\|_{L^2(\mathcal{D})} \leq cs|u|_H^m |h|_H^{n+1}$$

where c is a positive constant and $\xi \in \mathcal{D}, h \in H, s > 0$. Hence,

$$\lim_{s \to 0} \sup_{|h|_H \leq 1} \|G_1(\cdot, h, s)\|_{L^2(\mathcal{D})} = 0.$$

Similar arguments lead also to

$$\lim_{s \to 0} \sup_{|h|_H \leq 1} \|G_2(\cdot, h, s)\|_{L^2(\mathcal{D})} = 0.$$

Therefore,

$$\lim_{s \to 0} \sup_{|h|_H \leq 1} \|[(f' \circ (u + sh) - f' \circ (u)) \cdot h]^{(k)}\|_{L^2(\mathcal{D})} = 0, \quad r \leq k.$$

This completes the proof that $F : H \to H$ is Fréchet differentiable. The fact that F is r-times differentiable for $r \geq 2$ can be proved inductively using similar but lengthier computations. Details are left to the reader. □

Using Itô's formula, it is easy to see that the solution of the following H-valued linear stochastic differential equation

$$du^*(t) = Bu^*(t)\, dW(t), \quad u^*(0) = \psi \in H := H_0^k(\mathcal{D})$$

is given by

$$u^*(t, \psi, \omega)(\xi) := Q(t, \xi, \omega)\psi(\xi), \quad \xi \in \mathcal{D}, \psi \in H, t \geq 0,$$

where the process $Q : \mathbf{R}^+ \times \mathcal{D} \times \Omega \to \mathbf{R}$ is defined by

$$Q(t, \xi, \omega) := \exp\left\{\sum_{i=1}^\infty \sigma_i(\xi) W^i(t, \omega) - \frac{1}{2} \sum_{i=1}^\infty \sigma_i^2(\xi) t\right\},\ t \geq 0, \xi \in \mathcal{D}, \omega \in \Omega.$$

Using the perfect helix property of (W, θ), the reader may easily check the following cocycle identity for Q:

$$Q(t_1 + t_2, \xi, \omega) = Q(t_2, \xi, \theta(t_1, \omega)) Q(t_1, \xi, \omega) \quad t_1, t_2 \geq 0, \xi \in \mathcal{D}, \omega \in \Omega.$$

The above identity immediately implies that $u^* : \mathbf{R}^+ \times H \times \Omega \to H$ is a perfect linear cocycle with respect to the Brownian shift θ.

We now prove the following proposition:

PROPOSITION 1.3.4. *Assume $f \in C_b^k(R)$, $k > \frac{d}{2}$, and the forgoing conditions on the coefficients of the spde* (1.3.2). *Let S be a bounded subset of $H_0^k(\mathcal{D})$. Then for any $T > 0$ and almost all $\omega \in \Omega$, the weak solution $u(t,\psi)$ of the spde* (1.3.2) *satisfies*

$$\sup_{\psi \in S} \sup_{0 \le t \le a} \|u(t,\psi)\|_{H_0^k(\mathcal{D})} \le C(\omega, a),$$

for any $a \in \mathbf{R}^+$, where $C(\omega,a)$ is a random positive constant.

PROOF. Let $u(t,\psi)$ be the weak solution of the spde (1.3.2) with initial function $\psi \in H_0^k(\mathcal{D})$. Pick a sequence $\{\psi_n : n \ge 1\}$ of smooth functions in $C_b^\infty(\mathcal{D})$ such that $\psi_n \to \psi$ as $n \to \infty$ in $H_0^k(\mathcal{D})$. Let $u_n(t,\xi) := u(t,\psi_n)(\xi)$, $t \ge 0, \xi \in \mathcal{D}, n \ge 1$. Then each u_n, $n \ge 1$, is a strong solution of the spde (1.3.2). Define $v_n(t,\xi) := Q(t,\xi)^{-1} u_n(t,\xi)$, $t \ge 0, \xi \in \mathcal{D}$. Using the relations

$$dQ(t,\xi) = \sum_{i=1}^\infty \sigma_i(\xi) Q(t,\xi)\, dW^i(t), \quad t > 0, \xi \in \mathcal{D},$$

$$dQ(t,\xi)^{-1} = \sum_{i=1}^\infty \sigma_i^2(\xi) Q(t,\xi)^{-1}\, dt - \sum_{i=1}^\infty \sigma_i(\xi) Q(t,\xi)^{-1}\, dW^i(t), \quad t > 0, \xi \in \mathcal{D},$$

and Itô's formula, it follows that

$$\begin{aligned} dv_n(t,\xi) =& Q(t,\xi)^{-1} \frac{1}{2} \Delta u_n(t,\xi) dt + Q(t,\xi)^{-1} f(u_n(t,\xi)) dt \\ &+ Q(t,\xi)^{-1} u_n(t,\xi) \sum_{i=1}^\infty \sigma_i(\xi) dW^i(t) \\ &+ u_n(t,\xi) Q(t,\xi)^{-1} \sum_{i=1}^\infty \sigma_i^2(\xi) dt - u_n(t,\xi) Q(t,\xi)^{-1} \sum_{i=1}^\infty \sigma_i(\xi) dW^i(t) \\ &- u_n(t,\xi) Q(t,\xi)^{-1} \sum_{i=1}^\infty \sigma_i^2(\xi) dt \end{aligned}$$

a.s. for all $t > 0, \xi \in \mathcal{D}$,.

Therefore, for each $n \ge 1$, $v_n(t,\xi,\omega)$ satisfies the following parabolic equation with random coefficients:

(1.3.4^n)

$$\begin{aligned} \frac{\partial v_n}{\partial t} =& \frac{1}{2} \Delta v_n + < \nabla \ln Q(t,\xi), \nabla v_n >_{\mathbf{R}^d} - \Big[\frac{1}{2} Q(t,\xi) \Delta Q(t,\xi)^{-1} \\ &+ < \nabla Q(t,\xi), \nabla Q(t,\xi)^{-1} >_{\mathbf{R}^d} \Big] v_n + Q(t,\xi)^{-1} f(Q(t,\xi) v_n), \quad t > 0, \\ v_n(0,\xi) =& \psi_n(\xi). \end{aligned}$$

Let v denote the unique weak solution of the parabolic random pde

(1.3.4)

$$\begin{aligned} \frac{\partial v}{\partial t} =& \frac{1}{2} \Delta v + < \nabla \ln Q(t,\xi), \nabla v >_{\mathbf{R}^d} - \Big[\frac{1}{2} Q(t,\xi) \Delta Q(t,\xi)^{-1} \\ &+ < \nabla Q(t,\xi), \nabla Q(t,\xi)^{-1} >_{\mathbf{R}^d} \Big] v + Q(t,\xi)^{-1} f(Q(t,\xi) v), \quad t > 0, \\ v(0,\xi) =& \psi(\xi) \end{aligned}$$

for $t > 0, \xi \in \mathcal{D}$, with $\psi \in H_0^k(\mathcal{D})$. Since the coefficients of (1.3.4) are smooth, it is well known that

$$\lim_{n\to\infty} \sup_{0\le t\le a} \|v(t,\cdot,\omega) - v_n(t,\cdot,\omega)\|_{H_0^k(\mathcal{D})} = 0$$

for each $\omega \in \Omega$ and any $a \in \mathbf{R}^+$. By rewriting (1.3.4), it is easy to see that v satisfies the random pde

$$\frac{\partial v}{\partial t} = \frac{1}{2} Q(t,\xi)^{-1}\Delta(Q(t,\xi)v) + Q(t,\xi)^{-1} f(Q(t,\xi)v), \quad t > 0. \tag{1.3.5}$$

Since $\psi \in H_0^k(\mathcal{D}), k > \frac{d}{2}$, then by virtue of the Sobolev embedding of $H_0^k(\mathcal{D})$ into $L^\infty(\mathcal{D})$, we can view (1.3.4) as a random reaction diffusion equation in $L^\infty(\mathcal{D})$ whose non-linear term has linear growth and is globally Lipschitz. Now we use a standard argument to get an priori estimate for the solution of equation (1.3.4). Let $p(t,\eta,s,\xi)$ be the fundamental solution of the operator $\frac{1}{2}Q(t,\xi)^{-1}\Delta(Q(t,\xi)v)$, then it is well known that there are positive constants c_1, c_2 such that:

$$p(t,\eta,s,\xi) \le c_1(t-s)^{-\frac{d}{2}} \exp\left\{-\frac{c_2|\eta-\xi|^2}{2(t-s)}\right\},$$

for all $0 \le s < t \le a, \eta, \xi \in \mathcal{D}$ (see e.g. [Fri]). Therefore, it is easy to see that there is a positive constant c such that $\int_{\mathcal{D}} p(t,\eta,s,\xi)\,d\eta \le c$ for $0 \le s < t \le a, \xi \in \mathcal{D}$.

By the classical variation of parameters formula, the solution of (1.3.5) satisfies the random integral equation

$$\begin{aligned} v(t,\xi,\omega) = &\int_{\mathcal{D}} p(t,\eta,0,\xi)\psi(\eta)d\eta \\ &+ \int_0^t \int_{\mathcal{D}} p(t,\eta,s,\xi)Q(s,\eta,\omega)^{-1} f(Q(s,\eta,\omega)v(s,\eta,\omega))\,d\eta\,ds \end{aligned}$$

for $t > 0, \xi \in \mathcal{D}, \psi \in H_0^k(\mathcal{D}), \omega \in \Omega$. From the Lipschitz continuity of f, it is easy to see that

$$\begin{aligned} |v(t,\xi,\omega)| \le &\|\psi\|_\infty \int_{\mathcal{D}} p(t,\eta,0,\xi)\,d\eta \\ &+ \int_0^t \int_{\mathcal{D}} p(t,\eta,s,\xi)Q(s,\eta,\omega)^{-1}|f(Q(s,\eta,\omega)v(s,\eta,\omega)) - f(0)|\,d\eta\,ds \\ &+ \int_0^t \int_{\mathcal{D}} p(t,\eta,s,\xi)Q(s,\eta,\omega)^{-1}|f(0)|\,d\eta\,ds \\ \le &C(\|\psi\|_\infty + 1) + L\int_0^t \int_{\mathcal{D}} p(t,\eta,s,\xi)|v(s,\eta,\omega)|\,d\eta\,ds \\ \le &C_1(\|\psi\|_\infty + 1) + L\int_0^t \sup_{\eta\in\mathcal{D}}|v(s,\eta,\omega)| \int_{\mathcal{D}} p(t,\eta,s,\xi)d\eta\,ds \\ \le &C_1(\|\psi\|_\infty + 1) + L\int_0^t \sup_{\eta\in\mathcal{D}}|v(s,\eta,\omega)| \int_{\mathcal{D}} p(t,\eta,s,\xi)\,d\eta\,ds \\ \le &C_2(\|\psi\|_\infty + 1) + C_2\int_0^t \|v(s,\cdot,\omega)\|_\infty\,ds, \end{aligned}$$

for $0 \le t \le a$, $\omega \in \Omega$. In the above estimates, L is the Lipschitz constant of f and C_1, C_2 are positive constants. Hence, using Gronwall's inequality, it follows that

$$\sup_{t\in[0,a]} \|v(t,\cdot,\omega)\|_\infty < \infty \tag{1.3.6}$$

for each $\omega \in \Omega$. Needless to say, this bound depends on ω, but this does not affect our analysis here.

Now put $f \equiv 0$ in the random pde (1.3.5) and use uniqueness of solutions together with the identity

$$Q(t_1+t_2,\xi,\omega) = Q(t_2,\xi,\theta(t_1,\omega))Q(t_1,\xi,\omega) \quad t_1,t_2 \ge 0, \omega \in \Omega$$

in order to conclude that weak solutions of the linear spde

$$du(t,\xi) = \frac{1}{2}\Delta u(t,\xi)dt + Bu(t,\xi)dW(t), \quad u(0,\xi) = \psi(\xi) \tag{1.3.7}$$

yield a stochastic linear semiflow $\phi : \mathbf{R}^+ \times H_0^k(\mathcal{D}) \times \Omega \to H_0^k(\mathcal{D})$ such that (ϕ,θ) is a perfect $L(H_0^k(\mathcal{D}))$-valued cocycle. Full details of the argument are given in the proof of Theorem 1.4.1 in the next section.

It is easy to see that the weak solution u of the spde (1.3.2) satisfies the following random integral equation:

$$u(t,\xi,\omega) = \phi(t,\psi,\omega)(\xi) + \int_0^t \phi(t-s,\theta(s,\omega))F(u(s,\xi,\omega))\,ds, \tag{1.3.8}$$

for $t \ge 0, \psi \in H_0^k(\mathcal{D}), \xi \in \mathcal{D}$.

Now, using Lemma 1.3.2 together with (1.3.6), one gets a positive random constant C_k^1 such that

$$\|F(u(t,\cdot,\omega))\|_{H_0^k} \le C_k^1(\omega,a)\|u(t,\cdot,\omega)\|_{H_0^k}$$

for all $t \in [0,a], \omega \in \Omega$ and any $a \in \mathbf{R}^+$.

Finally, the assertion of the proposition follows from (1.3.8) and a simple application of Gronwall's lemma. □

THEOREM 1.3.5. *Suppose $k > \frac{d}{2}$. Assume $f : R \to R$ is a C_b^∞ function. Assume all the forgoing conditions on the coefficients and the noise term in the spde (1.3.2). Then for each $\psi \in H_0^k(\mathcal{D}))$ the spde (1.3.2) has a unique weak $(\mathcal{F}_t)_{t\ge0}$-adapted solution $u(\cdot,\psi,\cdot) : \mathbf{R}^+ \times \Omega \to H_0^k(\mathcal{D}))$. Furthermore, the family of weak solutions $u(\cdot,\psi,\cdot)$, $\psi \in H_0^k(\mathcal{D})$, of (1.3.2) admits a $(\mathcal{B}(\mathbf{R}^+)\otimes\mathcal{B}(H_0^k(\mathcal{D}))\otimes\mathcal{F},\mathcal{B}(H_0^k(\mathcal{D})))$-measurable version $u : \mathbf{R}^+ \times H_0^k(\mathcal{D}) \times \Omega \to H_0^k(\mathcal{D})$ having the following properties:*

(i) *For each $\psi \in H_0^k(\mathcal{D})$, $u(\cdot,\psi,\cdot) : \mathbf{R}^+ \times \Omega \to H_0^k(\mathcal{D})$ is $(\mathcal{F}_t)_{t\ge0}$-adapted.*
(ii) *(u,θ) is a C^∞ perfect cocycle on $H_0^k(\mathcal{D})$ (in the sense of* Definition 1.1.2).
(iii) *For each $(t,\omega) \in (0,\infty) \times \Omega$, the map $H_0^k(\mathcal{D}) \ni \psi \mapsto u(t,\psi,\omega) \in H_0^k(\mathcal{D})$ takes bounded sets into relatively compact sets.*
(iv) *For each $(t,\psi,\omega) \in (0,\infty) \times H_0^k(\mathcal{D}) \times \Omega$, and any integer $r \ge 1$, the Fréchet derivative $D^{(r)}u(t,\psi,\omega) \in L_2^{(r)}(H_0^k(\mathcal{D}),H_0^k(\mathcal{D}))$, and the map*

$$[0,\infty) \times H_0^k(\mathcal{D}) \times \Omega \ni (t,\psi,\omega) \mapsto D^{(r)}u(t,\psi,\omega) \in L^{(r)}(H_0^k(\mathcal{D}),H_0^k(\mathcal{D}))$$

is strongly measurable.

(v) *For any positive a, ρ and any positive integer r,*

$$E \log^+ \left\{ \sup_{\substack{0 \le t_1, t_2 \le a \\ \psi \in H_0^k(\mathcal{D})}} \frac{\|u(t_2, \psi, \theta(t_1, \cdot))\|_{H_0^k(\mathcal{D})}}{(1 + \|\psi\|_{H_0^k(\mathcal{D})})} \right\} < \infty$$

and

$$E \log^+ \sup_{\substack{0 \le t_1, t_2 \le a \\ \|\psi\|_{H_0^k(\mathcal{D})} \le \rho}} \{\|D^{(r)} u(t_2, \psi, \theta(t_1, \cdot))\|_{L^{(r)}(H_0^k(\mathcal{D}), H_0^k(\mathcal{D}))}\} < \infty.$$

PROOF. It is easy to see that the linear cocycle (ϕ, θ) of the spde (1.3.7) in the proof of Proposition 1.3.4 satisfies all the assertions in Theorem 1.2.1 and Theorem 1.2.2. In view of Proposition 1.3.4, the theorem now follows by a straightforward modification of the proof of Theorem 1.2.6 (See Remark (i) following the proof of Theorem 1.2.6). □

1.4. Semilinear spde's: Non-Lipschitz nonlinearity

In this section, we will study two types of semilinear spde's with non-Lipschitz nonlinearities and infinite dimensional noise.

The two classes of spde's considered are *stochastic reaction diffusion equations* and *stochastic Burgers equation with additive noise.* We prove the existence of a compacting C^1-cocycle in each case.

(a) *Stochastic reaction diffusion equations*

This class of spde's has dissipative nonlinear terms and infinite dimensional spatially smooth white noise. We prove the existence of a compacting $C^{0,1}$-cocycle satisfying appropriate regularity properties (Theorem 1.4.1). It appears that the cocycle is in general not Fréchet differentiable on the space of all L^2 functions on the domain (cf. [Te], p. 298). However, for a subclass of dissipative non-linearities with a certain dimension requirement, we further prove that the cocycle is C^1 and possesses Oseledec-type integrability properties (Theorem 1.4.2).

In [F.2], Flandoli studied the existence of *continuous* semi-flows for a class of spde's with finite dimensional noise and polynomial nonlinearities of odd degree and with negative leading coefficients.

Consider the following stochastic reaction diffusion equation in a smooth bounded domain $\mathcal{D} \subset \mathbf{R}^d$,

$$\left.\begin{aligned} du &= \nu \Delta u \, dt + f(u(t)) \, dt + \sum_{i=1}^{\infty} \sigma_i u \, dW^i(t), \quad t > 0 \\ u(0) &= \psi \\ u(t)|_{\partial \mathcal{D}} &= 0, \quad t > 0. \end{aligned}\right\} \tag{1.4.1}$$

where Δ is the Laplacian on $\mathcal{D}$, and $\nu > 0$ is a real constant. The initial function $\psi : \mathcal{D} \to \mathbf{R}$ is square-integrable with respect to Lebesgue measure on $\mathcal{D}$, and a Dirichlet boundary condition is assumed on the boundary $\partial \mathcal{D}$. The noise term $\sum_{i=1}^{\infty} \sigma_i u \, dW^i(t)$ is very similar to the one in (1.3.2) of section 1.3: In particular, $W^i, i \ge 1$, are standard one-dimensional Brownian motions on a complete filtered

Wiener space $(\Omega, \bar{\mathcal{F}}, (\mathcal{F}_t)_{t\geq 0}, P)$. The $\sigma_i : \mathcal{D} \to R$, $i \geq 1$, are assumed to be functions in the Sobolev space $H_0^s(\mathcal{D})$ with $s > 2 + \frac{d}{2}$, satisfying

$$\sum_{i=1}^{\infty} \|\sigma_i\|_{H_0^s}^2 < \infty.$$

The nonlinearity $f : \mathbf{R} \to \mathbf{R}$ satisfies the following classical dissipativity conditions:

Conditions (D):

The function f is C^2, and there are positive constants $c_i, i = 1, 2, 3, 4$, and a positive integer p such that

$$\begin{aligned} -c_2 - c_3 s^{2p} \leq f(s)s \leq c_2 - c_1 s^{2p} \\ f'(s) \leq c_4 \end{aligned}$$

for all $s \in \mathbf{R}$.

A typical example of a function $f : \mathbf{R} \to \mathbf{R}$ satisfying Conditions (D) is the polynomial $f(s) := \sum_{k=1}^{2p-1} a_k s^k$, $s \in \mathbf{R}$, where $a_{2p-1} < 0$. (See e.g. [Te], pp. 83-85.)

Solutions of (1.4.1) are to be understood in a *weak* sense as defined below.

Consider the Hilbert space $H := L^2(\mathcal{D})$ of all square-integrable functions $\psi : \mathcal{D} \to \mathbf{R}$ furnished with the L^2 inner product

$$< \psi_1, \psi_2 >_H := \int_{\mathcal{D}} \psi_1(\xi)\psi_2(\xi)\, d\xi, \quad \psi_1, \psi_2 \in H,$$

where $d\xi$ stands for Lebesgue measure on $\mathcal{D}$. Denote the induced norm on H by

$$|\psi|_H := \left[\int_{\mathcal{D}} |\psi(\xi)|^2\, d\xi\right]^{1/2}, \quad \psi \in H.$$

Recall $C_0^\infty(\mathcal{D})$, the set of all smooth test functions $\phi : \mathcal{D} \to \mathbf{R}$ which vanish on $\partial\mathcal{D}$. Let $L^\infty(\mathcal{D})$ stand for all essentially bounded measurable functions $\psi : \mathcal{D} \to \mathbf{R}$ with the usual norm

$$\|\psi\|_\infty := \text{essup}_{\xi\in\mathcal{D}} |\psi(\xi)|.$$

An $(\mathcal{F}_t)_{t\geq 0}$-adapted random field $u : \mathbf{R}^+ \times \mathcal{D} \times \Omega \to \mathbf{R}$ is a *weak solution* of (1.4.1) if $u(t, \cdot, \omega) \in H$ for a.a. $\omega \in \Omega, t > 0$, and the following identity holds:

$$\left.\begin{aligned} d < u(t), \phi >_H &= \nu < u(t), \Delta\phi >_H dt + < f(u(t)), \phi >_H dt \\ &\qquad\qquad + \sum_{i=1}^{\infty} < \sigma_i u(t), \phi >_H dW^i(t), \\ u(0) &= \psi \in L^2(\mathcal{D}), \\ u(t)|_{\partial\mathcal{D}} &= 0, \quad t > 0, \end{aligned}\right\}$$

for all $\phi \in C_0^\infty(\mathcal{D})$ a.s.

Note that, unless f has linear growth ($p = 1$ in Conditions (D)), the Nemytskii operator $F(u)(\xi) := f(u(\xi))$, $\xi \in \mathcal{D}$, does not even map $H = L^2(\mathcal{D})$ into itself. Thus one cannot view (1.4.1) as a semilinear see on H. Nevertheless, we will show that for each $\psi \in H$, (1.4.1) admits a unique weak solution $u(t) \in H$ a.s., for all $t > 0$. Furthermore, the ensemble of all weak solutions of (1.4.1) generates a globally Lipschitz cocycle (*also denoted by the same symbol*) $u : \mathbf{R}^+ \times H \times \Omega \to$

H satisfying the assertions of Theorem 1.4.1 below. In particular, the stochastic semiflow $u(t,\cdot,\omega) : H \to H$ takes bounded sets into relatively compact sets in H, and its global Lipschitz constant has moments of all orders.

As for Fréchet differentiability of the cocycle $u : \mathbf{R}^+ \times H \times \Omega \to H$ on the whole of H, it appears to be *not true* when f is smooth and satisfies Conditions (D) (cf. [Te], p. 298). However, under a stronger dimension requirement on the polynomial growth rate p of f, we are able to establish that the cocycle u is C^1 on H (Theorem 1.4.2). Furthermore, it satisfies similar assertions to those of Theorem 1.2.6. In particular, its Fréchet derivatives $Du(t,\psi,\omega) : H \to H$ are compact for all $(t,\psi,\omega) \in (0,\infty) \times H \times \Omega$.

In (1.4.1), the special case $f(s) := s(1-s)$, $s \in \mathbf{R}$, corresponds to the well-known stochastic KPP equation. It is not covered by the analysis in this section since it only admits positive solutions for all time. Its random travelling wave and ergodic properties were considered in [E-Z], [D-T-Z.1] and [O-V-Z]. For the KPP equation with additive noise, the reader may refer to [E-H] for the existence of the invariant measure.

The following lemma reduces (1.4.1) to a random family of reaction-diffusion equations.

LEMMA 1.4.1. *Recall the process* $Q : \mathbf{R}^+ \times \mathcal{D} \times \Omega \to \mathbf{R}$ *defined by* $Q(t,\xi,\omega) := \exp\left\{\sum_{i=1}^{\infty} \sigma_i(\xi) W^i(t,\omega) - \frac{1}{2}\sum_{i=1}^{\infty} \sigma_i^2(\xi) t\right\}$, $t \geq 0$, $\xi \in \mathcal{D}$, $\omega \in \Omega$. *Let* u *be a weak solution of* (1.4.1) *and set* $v(t,\xi,\omega) := Q(t,\xi,\omega)^{-1} u(t,\xi,\omega)$, $t \geq 0$, $\xi \in \mathcal{D}$, $\omega \in \Omega$. *Define* $\tilde{f} : \mathbf{R}^+ \times \mathcal{D} \times \mathbf{R} \times \Omega \to \mathbf{R}$ *by* $\tilde{f}(t,\xi,s,\omega) := Q(t,\xi,\omega)^{-1} f(Q(t,\xi,\omega)s)$, $t \in \mathbf{R}^+$, $\xi \in \mathcal{D}$, $s \in \mathbf{R}$, $\omega \in \Omega$. *Then* v *is a weak solution of the random reaction-diffusion equation*

$$
\left.\begin{aligned}
\frac{\partial v}{\partial t} &= \nu Q(t)^{-1} \Delta(Q(t)v) + \tilde{f}(t, v(t)), \quad t > 0\\
v(0) &= \psi \in L^2(\mathcal{D})\\
v(t)|_{\partial\mathcal{D}} &= 0, \quad t > 0.
\end{aligned}\right\} \tag{1.4.2}
$$

Conversely, every weak solution v *of* (1.4.2) *corresponds to a weak solution* u *of* (1.4.1) *given by* $u(t,\xi,\omega) := Q(t,\xi,\omega)v(t,\xi,\omega)$, $t \geq 0, \xi \in \mathcal{D}, \omega \in \Omega$.

PROOF. Suppose u is a weak solution of (1.4.1) with initial function $\psi \in L^2(\mathcal{D})$. Define

$$
v(t,\xi,\omega) := Q(t,\xi,\omega)^{-1} u(t,\xi,\omega),\ t \geq 0, \xi \in \mathcal{D}, \omega \in \Omega. \tag{1.4.3}
$$

Assume first that the initial function $\psi : \mathcal{D} \to \mathbf{R}$ is smooth. Then u is a strong solution of (1.4.1). Hence by Itô's formula (as in the proof of Proposition 1.3.4), it follows that v is a (strong) solution of the random reaction-diffusion equation (1.4.2). The case of a general $\psi \in L^2(\mathcal{D})$ can be handled by approximating ψ in the L^2-norm by a sequence of smooth functions $\psi_n : \mathcal{D} \to \mathbf{R}$, $n \geq 1$, as in the proof of Proposition 1.3.4.

A similar argument, using Itô's formula and the relation

$$
dQ(t,\xi) = \sum_{i=1}^{\infty} \sigma_i(\xi) Q(t,\xi)\, dW^i(t), \quad t > 0, \tag{1.4.4}
$$

proves the second assertion of the lemma. □

The next lemma shows that the non-linear term $\tilde{f}$ in (1.4.2) inherits the dissipativity properties of the original non-linear term f in (1.4.1).

LEMMA 1.4.2. *Suppose f satisfies* Conditions (D). *Let $0 < a < \infty$. Then there exist $\mathcal{F}$-measurable positive random variables $\tilde{c}_i \in \bigcap_{k=1}^{\infty} L^k(\Omega, \mathbf{R}), i = 1, 2, 3,$ such that the following is true:*

$$\left.\begin{aligned} -\tilde{c}_2(\omega) - \tilde{c}_3(\omega)s^{2p} \leq \tilde{f}(t,\xi,s,\omega)s \leq -\tilde{c}_1(\omega)s^{2p} + \tilde{c}_2(\omega), \\ \frac{\partial \tilde{f}(t,\xi,s,\omega)}{\partial s} \leq c_4 \end{aligned}\right\} \tag{1.4.5}$$

for all $t \in [0,a], s \in \mathbf{R}, \omega \in \Omega$.

PROOF. Fix $a \in (0,\infty), 0 \leq t \leq a, \xi \in \mathcal{D}, s \in \mathbf{R}, \omega \in \Omega$. Then Conditions (D) imply that

$$\begin{aligned} -c_2 Q(t,\xi,\omega)^{-2} - c_3 Q(t,\xi,\omega)^{(2p-2)} s^{2p} \leq& \tilde{f}(t,\xi,s,\omega)s \\ \leq& -c_1 Q(t,\xi,\omega)^{(2p-2)} s^{2p} + c_2 Q(t,\xi,\omega)^{-2}, \\ \frac{\partial \tilde{f}(t,s,\omega)}{\partial s} \leq& c_4. \end{aligned}$$

Define

$$\tilde{c}_1(\omega) := c_1 \inf_{0 \leq t \leq a, \xi \in \mathcal{D}} Q(t,\xi,\omega)^{(2p-2)}, \quad \tilde{c}_2(\omega) := c_2 \sup_{0 \leq t \leq a, \xi \in \mathcal{D}} Q(t,\xi,\omega)^{-2}$$
$$\tilde{c}_3(\omega) := c_3 \sup_{0 \leq t \leq a, \xi \in \mathcal{D}} Q(t,\xi,\omega)^{(2p-2)}$$

for all $\omega \in \Omega$. By sample continuity of $Q(t,\xi)$ and $Q(t,\xi)^{-1}$, it is clear that each $\tilde{c}_i(\omega), i = 1,2,3$, is finite for a.a. $\omega \in \Omega$. The estimates of the lemma follow immediately from the above inequalities and the definition of $Q(t,\xi)$. The existence of all moments of $\tilde{c}_i, i = 1,2,3$, follows from Burkholder-Davis-Gundy inequality and the fact that $Q(t,\xi)$ and $Q(t,\xi)^{-1}$ satisfy the linear sde's (1.4.3) and (1.4.4).□

In view of Lemmas 1.4.1 and 1.4.2, we can now adapt standard methods from deterministic pde's in order to prove Theorem 1.4.1 below. In particular, the existence of the stochastic semiflow for weak solutions of the spde (1.4.1) follows from the regularity properties of solutions to the random reaction diffusion equation (1.4.2). For the existence of the semiflow of (1.4.2), its global Lipschitz continuity and compactness, we refer the reader to [Te], pp. 80-102, 371-374. Note that Lemma 1.4.2 ensures that the non-linear time-dependent random term $\tilde{f}$ in (1.4.2) satisfies appropriate dissipativity estimates which carry sufficient uniformity in t to allow for the apriori estimates in [Te] to work. This renders the proof of Theorem 1.4.1 below an adaptation of the corresponding arguments in [Te]. Thus, we will only sketch the proof and leave many of the details to the reader.

THEOREM 1.4.1. *Assume that f in (1.4.1) satisfies* Conditions (D). *Then for each $\psi \in H := L^2(\mathcal{D})$, the spde (1.4.1) admits a unique $(\mathcal{F}_t)_{t\geq 0}$-adapted weak solution $u(\cdot,\psi,\cdot) : \mathbf{R}^+ \times \Omega \to H$ such that $u(\cdot,\psi,\omega) \in L^{2p}((0,T), L^{2p}(\mathcal{D})) \cap C(\mathbf{R}^+, H)$ for a.a. $\omega \in \Omega$. The family of all weak solutions of (1.4.1) has a $(\mathcal{B}(\mathbf{R}^+) \otimes \mathcal{B}(H) \otimes \mathcal{F}, \mathcal{B}(H))$-measurable version $u : \mathbf{R}^+ \times H \times \Omega \to H$ with the following properties:*

(i) *For each* $\psi \in H$, $u(\cdot,\psi,\cdot) : \mathbf{R}^+ \times \Omega \to H$ *is an* $(\mathcal{F}_t)_{t\geq 0}$*-adapted weak solution of* (1.4.1).
(ii) (u,θ) *is a* $C^{0,1}$ *perfect cocycle on* H *(in the sense of* Definition 1.1.2*).*
(iii) *For each* $(t,\omega) \in (0,\infty) \times \Omega$, *the map* $H \ni \psi \mapsto u(t,\psi,\omega) \in H$ *is globally Lipschitz and takes bounded sets in* H *into relatively compact sets.*
(iv) *For any positive* a, ρ,

$$E \log^+ \sup_{\substack{0\leq t\leq a \\ |\psi|_H \leq \rho}} |u(t,\psi,\cdot)|_H < \infty.$$

(v) *For each* $\omega \in \Omega$,

$$\limsup_{t\to\infty} \frac{1}{t} \log^+ \sup_{\substack{\psi_1 \neq \psi_2, \\ \psi_1,\psi_2 \in H}} \frac{|u(t,\psi_1,\omega) - u(t,\psi_2,\omega)|_H}{|\psi_1 - \psi_2|_H} \leq \frac{1}{2}\Big(c_4 - \nu\lambda_1 - \sigma^2\Big)$$

where $\sigma^2 := \inf_{\xi\in\mathcal{D}} \sum_{i=1}^{\infty} \sigma_i^2(\xi)$. *In particular, if*

$$\sup_{s\in\mathbf{R}} f'(s) - \nu\lambda_1 - \sigma^2 < 0,$$

then the stochastic flow $u(t,\cdot,\omega) : H \to H$ *is a uniform contraction for sufficiently large* $t > 0$.

PROOF. The existence and uniqueness of a weak solution of (1.4.1) follows from the corresponding result for the random reaction-diffusion equation (1.4.2) ([Te], pp. 89-91). Using the dissipativity estimates (1.4.5) on $\tilde{f}$, a straightforward modification of the Galerkin approximation technique in [Te] (pp. 89-91) gives the existence of a weak solution $u(\cdot,\psi,\omega) : \mathbf{R}^+ \to H$ of the random reaction-diffusion equation (1.4.2) for each fixed $\omega \in \Omega$ (cf. also [Ro], pp. 221-227). The joint measurability and $(\mathcal{F}_t)_{t\geq 0}$-adaptedness of the solution are also immediate consequences of the Galerkin approximations. This completes the proof of assertion (i) of the theorem.

To prove assertion (iii), denote by $v(\cdot,\psi,\cdot) : \mathbf{R}^+ \times \Omega \to H$, $\psi \in H$, the family of all weak solutions of the random pde (1.4.2). We will show that for each $\omega \in \Omega$, the map $H \ni \psi \mapsto u(t,\psi,\omega) \in H$ is globally Lipschitz uniformly in t over bounded sets in $\mathbf{R}^+$. To see this, let $\psi_i \in H, i = 1,2$. Denote by $v_i(t) := v(t,\psi_i), t \geq 0, i = 1,2$, the weak solutions of the random pde (1.4.2) starting at $\psi_i \in H, i = 1,2$. Then multiplying both sides of the equation

$$\frac{\partial(v_1(t) - v_2(t))}{\partial t} = \nu Q(t)^{-1}\Delta(Q(t)(v_1(t) - v_2(t))) + \tilde{f}(t,v_1(t)) - \tilde{f}(t,v_2(t)), \quad t > 0,$$

by $v_1(t) - v_2(t)$ and integrating over $\mathcal{D}$, we obtain

$$\begin{aligned}\int_{\mathcal{D}} (v_1(t) - v_2(t))\frac{\partial(v_1(t) - v_2(t))}{\partial t}\, d\xi \\ =&\nu < Q(t)^{-1}\Delta(Q(t)(v_1(t) - v_2(t))), v_1(t) - v_2(t) >_H \\ &+ \int_{\mathcal{D}} \tilde{f}(t,v_1(t)) - \tilde{f}(t,v_2(t))(v_1(t) - v_2(t))\, d\xi,\end{aligned}$$

for all $t > 0$. Using the Mean-Value Theorem and the second estimate in (1.4.5), it follows that

$$\begin{aligned}\frac{1}{2}\frac{d}{dt}|v_1(t) - v_2(t)|_H^2 =& \nu < Q(t)^{-1}\Delta(Q(t)(v_1(t) - v_2(t))), v_1(t) - v_2(t) >_H \\ &+ \int_{\mathcal{D}}\int_0^1 \frac{\partial \tilde{f}}{\partial s}(t, \lambda v_1(t) + (1-\lambda)v_2(t))\, d\lambda (v_1(t) - v_2(t))^2\, d\xi \\ \le& -\nu\lambda_1|v_1(t) - v_2(t)|_H^2 + c_4|v_1(t) - v_2(t)|_H^2 \qquad (1.4.6)\end{aligned}$$

for all $t > 0$. In the above inequality, λ_1 is the smallest eigenvalue of $-Q(t)^{-1}\Delta(Q(t)\cdot)$. This turns out to be the same as the smallest eigenvalue of $-\Delta$. Applying Gronwall's lemma to (1.4.6), we get

$$|v_1(t,\omega) - v_2(t,\omega)|_H^2 \le |\psi_1 - \psi_2|_H^2 \exp\{(c_4 - \nu\lambda_1)t\} \qquad (1.4.7)$$

for all $t \ge 0$ and all $\omega \in \Omega$. Using the relations $u(t, \psi_i, \omega) = Q(t, \xi, \omega)v_i(t, \omega)$, $i = 1, 2$, in (1.4.7), we deduce that

$$|u(t, \psi_1, \omega) - u(t, \psi_2, \omega)|_H \le |\psi_1 - \psi_2|_H \exp\left\{\frac{1}{2}\left(c_4 - \nu\lambda_1\right)t\right\} \sup_{\xi\in\mathcal{D}} Q(t, \xi, \omega)$$

for all $t \ge 0$, $\omega \in \Omega$, $\psi_1, \psi_2 \in H$. For any $a > 0$, define the random variable

$$c_5(\omega) := \sup_{0\le t\le a} \exp\left\{\frac{1}{2}\left(c_4 - \nu\lambda_1\right)t\right\} \sup_{\xi\in\mathcal{D}} Q(t, \xi, \omega), \quad \omega \in \Omega.$$

Then it is easy to see that $E\log^+ c_5 < \infty$, and

$$|u(t, \psi_1, \omega) - u(t, \psi_2, \omega)|_H \le c_5(\omega)|\psi_1 - \psi_2|_H \qquad (1.4.8)$$

for all $t \in [0, a]$, $\omega \in \Omega$, $\psi_1, \psi_2 \in H$. This proves the first assertion in (iii). (Note that (1.4.8) implies pathwise uniqueness of the weak solution to the spde (1.4.1): Just put $\psi_1 = \psi_2 = \psi$, a given initial function in H.) The local compactness of the semiflow $H \ni \psi \mapsto u(t, \psi, \omega) \in H$, $t > 0, \omega$, follows from the fact that the map $H \ni \psi \mapsto v(t, \psi, \omega) \in H$, $t > 0, \omega$, takes bounded sets in H to relatively compact sets.

We next prove the perfect cocycle property in (ii). To this end, fix $\psi \in H, \omega \in \Omega, t_1, t_2 \ge 0$. Define

$$Y(t) := v(t + t_1, \psi, \omega), \quad Z(t) := Q(t_1, \omega)^{-1}v(t, Q(t_1, \omega)v(t_1, \psi, \omega), \theta(t_1, \omega))$$

for all $t \ge 0$. Recall the perfect cocycle identity:

$$Q(t_1 + t_2, \xi, \omega) = Q(t_2, \xi, \theta(t_1, \omega))Q(t_1, \xi, \omega) \quad t_1, t_2 \ge 0, \xi \in \mathcal{D}. \qquad (1.4.9)$$

By the definition of $\tilde{f}$ in Lemma 1.4.1 and the above cocycle property, one gets

$$\begin{aligned}\tilde{f}(t, \xi, s, \theta(t_1, \omega)) =& Q(t, \xi, \theta(t_1, \omega))^{-1} f(Q(t, \xi, \theta(t_1, \omega))s) \\ =& Q(t_1, \xi, \omega)Q(t + t_1, \xi, \omega)^{-1} f(Q(t + t_1, \xi, \omega)Q(t_1, \xi, \omega)^{-1}s) \\ =& Q(t_1, \xi, \omega)\tilde{f}(t + t_1, \xi, Q(t_1, \xi, \omega)^{-1}s, \omega) \qquad (1.4.10)\end{aligned}$$

for all $s \in \mathbf{R}, t \ge 0, \xi \in \mathcal{D}$.

We now claim that the weak solution of the random reaction-diffusion equation (1.4.2) satisfies the following identity

$$v(t + t_1, \psi, \omega) = Q(t_1, \omega)^{-1}v(t, Q(t_1, \omega)v(t_1, \psi, \omega), \theta(t_1, \omega)) \qquad (1.4.11)$$

for all $t \geq 1$. This says that $Y(t) = Z(t)$ for all $t \geq 0$. Using (1.4.11) and the relation between u and v, it is easy to check that (u, θ) is a perfect cocycle. So we need only prove (1.4.11). By the definition of Z and (1.4.10), it follows that

$$\begin{aligned}\frac{\partial Z}{\partial t} =& \nu Q(t_1,\omega)^{-1}Q(t,\theta(t_1,\omega))^{-1}\Delta(Q(t,\theta(t_1,\omega))v(t,Q(t_1,\omega)v(t_1,\psi,\omega),\theta(t_1,\omega))) \\ &+ Q(t_1,\omega)^{-1}\tilde{f}(t,v(t,Q(t_1,\omega)v(t_1,\psi,\omega),\theta(t_1,\omega)),\theta(t_1,\omega)) \\ =& \nu Q(t+t_1,\omega)^{-1}\Delta(Q(t+t_1,\omega)Z(t)) \\ &+ \tilde{f}(t+t_1,Q(t_1,\omega)^{-1}v(t,Q(t_1,\omega)v(t_1,\psi,\omega),\theta(t_1,\omega)),\omega) \\ =& \nu Q(t+t_1,\omega)^{-1}\Delta(Q(t+t_1,\omega)Z(t)) + \tilde{f}(t+t_1,Z(t),\omega), \quad t>0;\end{aligned}$$

and $Z(0) = v(t_1, \psi, \omega)$. Now from its definition, Y also satisfies the same random pde:

$$\frac{\partial Y}{\partial t} = \nu Q(t+t_1,\omega)^{-1}\Delta(Q(t+t_1,\omega)Y(t)) + \tilde{f}(t+t_1,Y(t),\omega), \quad t>0$$

with the same initial condition $Y(0) = v(t_1, \psi, \omega)$. Therefore, by uniqueness of weak solutions to the above pde, we must have $Y(t) = Z(t)$ for all $t \geq 0$. This proves our claim, and hence (u, θ) is a perfect cocycle on H.

Assertion (v) of the theorem follows easily from (1.4.8). This completes the proof of the theorem. □

Our next result establishes Fréchet differentiability of the cocycle generated by the stochastic reaction diffusion equation:

$$\left.\begin{aligned} du =& \nu\Delta u\,dt + (1-|u|^\alpha)u\,dt + \sum_{i=1}^{\infty}\sigma_i(\xi)u\,dW^i(t), \quad t>0 \\ u(0) =& \psi \in H := L^2(\mathcal{D}) \\ u(t)|_{\partial\mathcal{D}} =& 0, \quad t>0. \end{aligned}\right\} \tag{1.4.12}$$

where $\nu > 0$ is a positive constant, and Δ is the Laplacian on a smooth bounded domain $\mathcal{D}$ with Dirichlet boundary conditions. As before, W^i, $i \geq 1$, are independent standard one-dimensional Brownian motions on the complete filtered Wiener space $(\Omega, \bar{\mathcal{F}}, (\mathcal{F}_t)_{t\geq 0}, P)$. Fréchet differentiability of the cocycle is established under the dimension requirement $\alpha < \frac{4}{d}$. It is not clear whether this condition is *necessary* for Fréchet differentiability of the cocycle.

THEOREM 1.4.2. *In (1.4.12), assume that $\alpha < \frac{4}{d}$. Then for each $\psi \in H := L^2(\mathcal{D})$, the spde (1.4.12) admits a unique $(\mathcal{F}_t)_{t\geq 0}$-adapted weak solution $u(\cdot,\psi,\cdot) : \mathbf{R}^+ \times \Omega \to H$ such that $u(\cdot,\psi,\omega) \in L^{2p}((0,T), L^{2p}(\mathcal{D})) \cap C(\mathbf{R}^+, H)$ for a.a. $\omega \in \Omega$. The family of all weak solutions of (1.4.12) has a $(\mathcal{B}(R^+) \otimes \mathcal{B}(H) \otimes \mathcal{F}, \mathcal{B}(H))$-measurable version $u : \mathbf{R}^+ \times H \times \Omega \to H$ with the following properties:*
(i) *For each $\psi \in H$, $u(\cdot,\psi,\cdot) : \mathbf{R}^+ \times \Omega \to H$ is an $(\mathcal{F}_t)_{t\geq 0}$-adapted weak solution of (1.4.12).*
(ii) *(u, θ) is a C^1 perfect cocycle on H (in the sense of* Definition 1.1.2*).*
(iii) *For each $(t, \omega) \in (0, \infty) \times \Omega$, the map $H \ni \psi \mapsto u(t,\psi,\omega) \in H$ is globally Lipschitz and takes bounded sets in H into relatively compact sets.*

(iv) *For each* $(t, \psi, \omega) \in (0, \infty) \times H \times \Omega$, *the Fréchet derivative* $Du(t, \psi, \omega) \in L(H)$ *is compact, and the map*

$$[0, \infty) \times H \times \Omega \ni (t, \psi, \omega) \mapsto Du(t, \psi, \omega) \in L(H)$$

is strongly measurable.

(v) *For any positive* a, ρ,

$$E \log^{+} \sup_{\substack{0 \le t \le a \\ |\psi|_H \le \rho}} \left\{ |u(t, \psi, \cdot)|_H + \|Du(t, \psi, \cdot)\|_{L(H)} \right\} < \infty.$$

PROOF. Fix any $\psi \in H = L^2(\mathcal{D})$. The existence and uniqueness of the solution to (1.4.12) in $L^2(\mathcal{D})$ is well-known as the nonlinear term satisfies the dissipativity condition ([D-Z.1]). This also follows by a similar argument to the proof of Theorem 1.4.1. So assertion (i) follows easily. The main purpose is to prove assertions (ii), (iii) and (iv). Recall that $Q(t, \xi) := \exp\left\{ \sum_{i=1}^{\infty} \sigma_i(\xi) W^i(t) - \frac{1}{2} \sum_{i=1}^{\infty} \sigma_i(\xi)^2 t \right\}$, $t \ge 0, \xi \in \mathcal{D}$, and let $v(t) = u(t) Q^{-1}(t), t \ge 0$.

For simplicity of notation and till further notice, we will suppress the dependence of the random fields u, v, etc. on ω.

Observe that $v(t, \psi)$ is a weak solution of the random reaction diffusion equation

$$\frac{\partial v}{\partial t} = \nu \Delta v + 2\nu Q(t)^{-1} \nabla Q(t) \nabla v + v(\nu Q(t)^{-1} \Delta Q(t) + 1 - Q^{\alpha}(t) |v|^{\alpha}) \tag{1.4.13}$$

for $t > 0$. By the Feynman-Kac formula, we have

(1.4.14)

$$v(t, \psi)(\xi) = \hat{E}[\chi_{\tau_t = t} \psi(x_t) e^{\int_0^t (\nu Q(t-s, x_s)^{-1} \Delta Q(t-s, x_s) + 1 - Q^{\alpha}(t-s, x_s) |v|^{\alpha}(t-s, \psi)(x_s)) ds}]$$

where x is the solution of the following stochastic differential equation

$$dx(s) = \sqrt{2\nu} dB(s) + 2\nu \nabla \log Q(t-s, x_s) ds, \quad x_0 = \xi \in \mathcal{D}, \ 0 < s < t,$$

and B is a Brownian motion in $\mathbf{R}^d$ independent of the $W^i, i \ge 1$. In (1.4.14), $\tau_t := \min(\tau, t)$, where τ is the first time the diffusion x hits $\partial \mathcal{D}$. Define $\beta := \nu \sup_{0 \le s \le t \le a, \xi \in \mathcal{D}} \dfrac{\Delta Q(t-s, \xi)}{Q(t-s, \xi)}$ for any $a > 0$. It follows from Jensen's inequality and (1.4.14) that

$$\begin{aligned} |v(t, \psi)|_H^2 \le & \int_{\mathcal{D}} \left(\hat{E} \chi_{\tau_t = t} |\psi(x_t)| e^{(\beta+1)t} \right)^2 d\xi \\ \le & e^{2(\beta+1)t} \int_{\mathcal{D}} \left(\int_{\mathcal{D}} p(t, \xi, y) |\psi(y)| dy \right)^2 d\xi \\ \le & e^{2(\beta+1)t} \int_{\mathcal{D}} \int_{\mathcal{D}} p(t, \xi, y) (\psi(y))^2 dy d\xi \\ \le & e^{2(\beta+1)t} |\psi|_H^2, \qquad 0 \le t \le a. \end{aligned} \tag{1.4.15}$$

In the above inequalities, $p(t, \xi, y)$ denotes the heat kernel associated with $\nu \Delta + 2\nu (\nabla \log Q(t)) \nabla$ on $\mathcal{D}$ with Dirichlet boundary condition. Define the induced heat semigroup $T_t : H \to H, t \geq 0$, by

$$(T_t\psi)(\xi) := \int_{\mathcal{D}} p(t,\xi,y)\psi(y)dy, \quad \psi \in H, \xi \in \mathcal{D}, t \geq 0.$$

Note that there exists a constant $c > 0$ such that

$$p(t,\xi,y) \leq \frac{c}{t^{\frac{d}{2}}}, \quad \xi, y \in \mathcal{D}, t > 0. \tag{1.4.16}$$

It is easy to see, using Jensen's inequality and (1.4.16), that

$$\begin{aligned}
|v(t,\psi)(\xi)|^2 \leq & \left(\ddot{E}\chi_{\tau_t=t}|\psi(x_t)|e^{(\beta+1)t} \right)^2 \\
\leq & e^{2(\beta+1)t}\hat{E}\chi_{\tau_t=t}(\psi(x_t))^2 \\
\leq & e^{2(\beta+1)t}\int_{\mathcal{D}} p(t,\xi,y)(\psi(y))^2 dy \\
\leq & e^{2(\beta+1)t}\frac{c}{t^{\frac{d}{2}}}\int_{\mathcal{D}} \psi^2(y)\, dy
\end{aligned}$$

for all $\xi \in \mathcal{D}$ and $t > 0$. Hence

$$\|v(t,\psi)\|_\infty \leq \frac{\sqrt{c}\, e^{(\beta+1)t}}{t^{\frac{d}{4}}}|\psi|_H \tag{1.4.17}$$

for $t > 0$.

Now let $\psi, g \in L^2(\mathcal{D})$ with $|g|_H \leq 1$, and h be a small real number. Since v is a mild solution of (1.4.13), it follows that $v(t,\psi+hg) - v(t,\psi)$ satisfies the following convolution equation in H:

(1.4.18)

$$\begin{aligned}
v(t,&\psi+hg) - v(t,\psi) \\
&= hT_t g + \int_0^t T_{t-s}(v(s,\psi+hg) - v(s,\psi))(1 + \frac{\Delta Q(s)}{Q(s)})ds \\
&\quad + \int_0^t T_{t-s}Q^\alpha(s)[v(s,\psi)|v(s,\psi)|^\alpha - v(s,\psi+hg)|v(s,\psi+hg)|^\alpha]ds, t > 0.
\end{aligned}$$

Define $m(x) := x|x|^\alpha$ for each $x \in \mathbf{R}$. Then $m'(x) = (\alpha+1)|x|^\alpha$, $x \in \mathbf{R}$. By the Mean-Value Theorem, we have

(1.4.19)

$$\begin{aligned}
&v(s,\psi+hg)|v(s,\psi+hg)|^\alpha - v(s,\psi)|v(s,\psi)|^\alpha \\
&\quad = (\alpha+1)\int_0^1 |rv(s,\psi+hg) + (1-r)v(s,\psi)|^\alpha dr(v(s,\psi+hg) - v(s,\psi))
\end{aligned}$$

for all $s \in \mathbf{R}^+$. Combining (1.4.18) and (1.4.19), we obtain

$$\begin{aligned}
&|v(t,\psi+hg) - v(t,\psi)|_H \\
&\quad \leq h|g|_H + c_1(\omega)\int_0^t |v(s,\psi+hg) - v(s,\psi)|_H\, ds \\
&\quad\quad + c_1(\omega)(\alpha+1)\int_0^t (\|v(s,\psi+hg)\|_{L^\infty} + \|v(s,\psi)\|_{L^\infty})^\alpha \times \\
&\quad\quad\quad \times |v(s,\psi+hg) - v(s,\psi)|_H\, ds,
\end{aligned}$$

for all $0 \leq t \leq a$, where $c_1(\omega)$ is a positive random constant. By virtue of (1.4.17), we get
(1.4.20)

$$\begin{aligned}|v(t,\psi+hg)-v(t,\psi)|_H \leq & h|g|_H + c_1(\omega)\int_0^t |v(s,\psi+hg)-v(s,\psi)|_H\,ds \\ & + c_2(\omega)(\alpha+1)\int_0^t \frac{1}{s^{\frac{d}{4}\alpha}}|v(s,\psi+hg)-v(s,\psi)|_H\,ds,\end{aligned}$$

for all $0 \leq t \leq a$, where $c_2(\omega) > 0$ is a random constant depending on the ball $\{g \in H : |g|_H \leq 1\}$. Using Gronwall's lemma and the requirement $\alpha d < 4$, we obtain

$$\sup_{0\leq t\leq a}\ \sup_{\substack{g\in H\\ |g|_H\leq 1}} |v(t,\psi+hg)-v(t,\psi)|_H \leq C(\omega)h.$$

Consider the $L(H)$-valued integral equation:
(1.4.21)

$$G_t(\psi) = T_t + \int_0^t T_{t-s}\Big(Q(s)^{-1}\Delta Q(s) + 1 - (\alpha+1)Q^\alpha(s)|v(s,\psi)|^\alpha\Big)G_s(\psi)ds,\ t\geq 0,$$

where $1-(\alpha+1)Q^\alpha(s)|v(s,\psi)|^\alpha$ is regarded as a multiplication operator on $L^2(\mathcal{D})$ whose operator norm satisfies the inequality

$$\|Q(s)^{-1}\Delta Q(s) + 1 - (\alpha+1)Q^\alpha(s)|v(s,\psi)|^\alpha\|_\infty \leq C_a(\omega)\frac{1}{s^{\frac{d}{4}\alpha}}|\psi|_H^\alpha, \tag{1.4.22}$$

for all $0 < s \leq a$, and for a positive random constant $C_a(\omega)$.

Claim: There exists a unique, continuous solution $[0,\infty) \ni t \mapsto G_t(\psi) \in L(H)$ to equation (1.4.21). Moreover for $t > 0$, $G_t(\psi) : H \to H$ is compact.

Proof of claim: Let $G_t^1(\psi) = T_t, t \geq 0$. Define the sequence $\{G_t^n(\psi)\}_{n=1}^\infty$ inductively by
(1.4.23)

$$G_t^{n+1}(\psi) = T_t + \int_0^t T_{t-s}\Big(Q(s)^{-1}\Delta Q(s) + 1 - (\alpha+1)Q^\alpha(s)|v(s,\psi)|^\alpha\Big)G_s^n(\psi)\,ds$$

for $n \geq 1$. Then (1.4.22) and (1.4.23) imply that

$$\|G_t^{n+1}(\psi)\|_{L(H)} \leq 1 + C_a(\omega)\|\psi\|^\alpha \int_0^t \frac{1}{s^{\frac{d}{4}\alpha}}\|G_s^n(\psi)\|_{L(H)}ds.,\quad 0 < t \leq a \tag{1.4.24}$$

Since $\frac{\alpha d}{4} < 1$, then by the standard successive approximation technique it follows that the sequence $\{G_\cdot^n(\psi)\}_{n=1}^\infty$ converges to the unique solution of (1.4.21). Next we prove that $G_t(\psi)$ is compact for each $t > 0$. It suffices to show that a Cauchy sequence can be extracted from the set $\{G_t(\psi)(g) : |g|_H \leq 1\}$ for each $t > 0$. Let $\delta_m, m \geq 1$, be a sequence of positive numbers decreasing to zero. Since T_t is compact for every $t > 0$, by a diagonal process there exists a sequence $g_n \in H$ with $|g_n|_H \leq 1$ such that $T_{\delta_m}g_n, n \geq 1$ is a Cauchy sequence for every m. Since $T_t, t \geq 0$, is a contraction semigroup on H, it is easy to see that $T_t g_n, n \geq 1$ is a

Cauchy sequence for every $t > 0$. Now consider

$$(1.4.25)\quad \begin{aligned} G_t(\psi)(g_n) - G_t(\psi)(g_m) &= T_t(g_n - g_m) \\ &+ \int_0^t T_{t-s}\Big(Q(s)^{-1}\Delta Q(s) + 1 - (\alpha+1)Q^\alpha(s)|v(s,\psi)|^\alpha\Big) \\ &\qquad \cdot (G_s(\psi)(g_n) - G_s(\psi)(g_m))ds, \quad t \ge 0. \end{aligned}$$

Hence,

(1.4.26)
$$\begin{aligned} &|G_t(\psi)(g_n) - G_t(\psi)(g_m)|_H \\ &\le |T_t g_n - T_t g_m|_H + C_a(\omega)|\psi|_H^\alpha \int_0^t \frac{1}{s^{\frac{d}{4}\alpha}} |G_s(\psi)(g_n) - G_s(\psi)(g_m)|_H \, ds. \end{aligned}$$

for all $t \in [0,a]$. Set $l(t) := \limsup_{n,m\to\infty} |G_t(\psi)(g_n) - G_t(\psi)(g_m)|_H$, $0 \le t \le a$. Taking $\limsup_{n,m\to\infty}$ on both sides of (1.4.26) we obtain

$$l(t) \le C_a(\omega)|\psi|_H^\alpha \int_0^t \frac{1}{s^{\frac{d}{4}\alpha}} l(s)ds, \quad 0 < t \le a.$$

This implies that $l(t) = 0$ for all $t > 0$, and completes the proof of the claim.

Next we show that v is Fréchet differentiable and $Dv(t,\psi) = G_t(\psi)$ for all $t \ge 0$. First we note that by using the Feynman-Kac formula and a similar argument as in the proof of (1.4.15) and (1.4.17), one has

$$(1.4.27)\quad \int_{\mathcal{D}} G_t(\psi)(g)(\xi)d\xi \le e^{2(\beta+1)t}|g|_H^2,$$

and

$$(1.4.28)\quad G_t(\psi)(g)(\xi) \le \frac{\sqrt{c}\, e^{(\beta+1)t}}{t^{\frac{d}{4}}} |g|_H^2, \quad 0 < t \le a.$$

Denote

$$(1.4.29)\quad \mu_t(\psi, g) := \frac{1}{h}(v(t,\psi+hg) - v(t,\psi)) - G_t(\psi)(g), \quad t > 0.$$

It is easy to see that μ satisfies the following integral equation:

$$(1.4.30)\quad \begin{aligned} \mu_t(\psi,g) &= \int_0^t T_{t-s}\Big(1 + \frac{\Delta Q(s)}{Q(s)}\Big)\mu_s(\psi,g)ds \\ &- \int_0^t T_{t-s}Q^\alpha(s)\Big[\frac{1}{h}(v(s,\psi+hg)|v(s,\psi+hg)|^\alpha - v(s,\psi)|v(s,\psi)|^\alpha)\Big]ds \\ &+ (\alpha+1)\int_0^t T_{t-s}Q^\alpha(s)[|v(s,\psi)|^\alpha G_s(\psi)(g)]ds, \quad t \ge 0. \end{aligned}$$

Using $m(y) - m(x) = \int_0^1 m'(ry + (1-r)x)dr(y-x)$ it follows from (1.4.30) that
(1.4.31)

$$\begin{aligned}\mu_t(\psi, g) = &\int_0^t T_{t-s}\left(1 + \frac{\Delta Q(s)}{Q(s)}\right)\mu_s(\psi, g)ds \\ &- (\alpha+1)\int_0^t T_{t-s}Q^\alpha(s)\left[\int_0^1 |rv(s,\psi+hg) \right. \\ &\left. \qquad + (1-r)v(s,\psi)|^\alpha dr\, \mu_s(\psi,g)\right] ds \\ &+ (\alpha+1)\int_0^t T_{t-s}Q^\alpha(s)\left[\int_0^1 (|v(s,\psi)|^\alpha - |rv(s,\psi+hg) \right. \\ &\left. \qquad + (1-r)v(s,\psi)|^\alpha)dr\, G_s(\psi)(g)\right] ds, \quad t \geq 0.\end{aligned}$$

Set $D(t) := \sup_{|g|_H \leq 1} |\mu_t(\psi, g)|_H$, $t \geq 0$. Using the L^∞ bound on $v(s, \psi + hg)$ this implies that for $0 < t \leq a$, one has
(1.4.32)

$$\begin{aligned}D(t) \leq &C(\omega)\int_0^t D(s)ds + C(\omega)\int_0^t \frac{1}{s^{\frac{d}{4}\alpha}} D(s)ds \\ &+ C(\omega)\sup_{|g|_H\leq 1}\int_0^t \left|\int_0^1 (|v(s,\psi)|^\alpha \right. \\ &\left. \qquad - |rv(s,\psi+hg) + (1-r)v(s,\psi)|^\alpha)\, dr G_s(\psi)(g)\right|_H ds.\end{aligned}$$

Again by Gronwall's lemma, it follows that there is a random constant $C(\omega)$ such that

$$\begin{aligned}D(t) \leq C(\omega)\sup_{|g|_H\leq 1}\int_0^t &\left|\left(\int_0^1 (|v(s,\psi)|^\alpha - |rv(s,\psi+hg)\right.\right. \\ &\left.\left. +(1-r)v(s,\psi)|^\alpha)dr\right)G_s(\psi)(g)\right|_H ds\end{aligned} \tag{1.4.33}$$

for all $t \in [0, a]$. To complete the proof of assertions (ii) and (iv) of the theorem, it suffices to show that
(1.4.34)

$$\lim_{h\to 0}\sup_{|g|_H\leq 1}\int_0^t |(\int_0^1 (|v(s,\psi)|^\alpha - |rv(s,\psi+hg) + (1-r)v(s,\psi)|^\alpha)dr)G_s(\psi)(g)|_H\, ds = 0$$

for all $t \in [0, a]$. Let us prove (1.4.34) for $\alpha \leq 1$ and $\alpha > 1$ separately. Assume first $\alpha \leq 1$. By Hölder inequality,

$$\begin{aligned}
&\sup_{|g|_H \leq 1} \int_0^t |(\int_0^1 (|v(s,\psi)|^\alpha - |rv(s,\psi+hg) + (1-r)v(s,\psi)|^\alpha) dr) G_s(\psi)(g)|_H \, ds \\
\leq &\sup_{|g|_H \leq 1} \int_0^t |(|v(s,\psi) - v(s,\psi+hg)|^\alpha) G_s(\psi)(g)|_H \, ds \\
\leq &\sup_{|g|_H \leq 1} \int_0^t \|G_s(\psi)(g)\|_{L^\infty}^\alpha (|v(s,\psi) - v(s,\psi+hg)|_H^\alpha) |G_s(\psi)(g)|^{1-\alpha} \| ds \\
\leq &\sup_{|g|_H \leq 1} \int_0^t \|G_s(\psi)(g)\|_{L^\infty}^\alpha |v(s,\psi) - v(s,\psi+hg)|_H^\alpha |G_s(\psi)(g)|_H^{1-\alpha} \, ds.
\end{aligned}$$

By virtue of (1.4.20) and (1.4.28), we get

$$\begin{aligned}
&\sup_{|g|_H \leq 1} \int_0^t |(\int_0^1 (|v(s,\psi)|^\alpha - |rv(s,\psi+hg) + (1-r)v(s,\psi)|^\alpha) dr) G_s(\psi)(g)\| ds \\
&\qquad \leq C_\alpha(\omega) h^\alpha \int_0^t \frac{1}{s^{\frac{d}{4}\alpha}} ds = C_\alpha(\omega) \frac{1}{1 - \frac{\alpha d}{4}} t^{1-\frac{\alpha d}{4}} h^\alpha, \quad 0 < t \leq a,
\end{aligned}$$

where $C_\alpha(\omega)$ is a random constant depending on the set $\{g \in L^2(\mathcal{D}) : \|g\|_{L^2} \leq 1\}$. This implies (1.4.34).

Assume now that $\alpha > 1$. Then

(1.4.35)

$$\begin{aligned}
&\int_0^t \left| \left(\int_0^1 (|v(s,\psi)|^\alpha - |rv(s,\psi+hg) + (1-r)v(s,\psi)|^\alpha) dr \right) G_s(\psi)(g) \right|_H ds \\
\leq &\int_0^t ds \int_0^1 dr |(|v(s,\psi)|^\alpha - |rv(s,\psi+hg) + (1-r)v(s,\psi)|^\alpha) G_s(\psi)(g)|_H \\
\leq &\int_0^t ds \int_0^1 dr \int_0^1 dk \alpha |(k|v(s,\psi)| + (1-k)|rv(s,\psi+hg) + (1-r)v(s,\psi)|)^{\alpha-1} \\
&\qquad \cdot |v(s,\psi+hg) - v(s,\psi)| G_s(\psi)(g)|_H \\
\leq &\int_0^t ds \int_0^1 dr \int_0^1 dk \alpha \|(k|v(s,\psi)| + (1-k)|rv(s,\psi+hg) + (1-r)v(s,\psi)|)^{\alpha-1}\|_{L^\infty} \\
&\qquad \cdot \|G_s(\psi)(g)\|_{L^\infty} |v(s,\psi+hg) - v(s,\psi)|_H, \quad 0 < t \leq a.
\end{aligned}$$

By (1.4.17), (1.4.20) and (1.4.28) it follows from (1.4.35) that

(1.4.36)

$$\begin{aligned}
&\sup_{|g|_H \leq 1} \int_0^t \left| \left(\int_0^1 (|v(s,\psi)|^\alpha - |rv(s,\psi+hg) + (1-r)v(s,\psi)|^\alpha) \, dr \right) G_s(\psi)(g) \right|_H ds \\
&\qquad \leq C_\alpha(\omega) h \int_0^t \frac{1}{s^{\frac{d}{4}(\alpha-1)}} \frac{1}{s^{\frac{d}{4}}} ds, \quad 0 < t \leq a.
\end{aligned}$$

This implies (1.4.34). So assertion (iv) holds.

To establish assertion (iii) of the theorem, use (1.4.18), (1.4.19) and a similar argument to the proof of (1.4.20), to obtain the following inequality
(1.4.37)

$$|v(t,\psi_m)-v(t,\psi_n)|_H \le |T_t\psi_m - T_t\psi_n|_H + C_1(\omega)\int_0^t |v(s,\psi_m)-v(s,\psi_n)|_H\,ds$$
$$+ C_2(\omega)(\alpha+1)\int_0^t \frac{1}{s^{\frac{d}{4}\alpha}}|v(s,\psi_m)-v(s,\psi_n)|_H\,ds, \quad 0<t\le a,$$

for $\psi_n,\psi_m \in H$ such that $|\psi_m|_H, |\psi_n|_H \le 1$. As in the proof of the compactness of $Dv(t,\psi)$, we can select a subsequence denoted also by $\{\psi_n\} \subset \{\psi : |\psi|_H \le 1\}$ such that for each $t>0$, $|T_t\psi_n - T_t\psi_m|_H \to 0$ as $n,m\to\infty$. One then can prove from (1.4.37) that $\lim_{n,m\to\infty} |v(t,\psi_n)-v(t,\psi_m)|_H = 0$. Therefore $v(t,\cdot): H \to H$ is compact or each $t>0$. This implies the compactness of each Fréchet derivative $Du(t,\psi,\omega): H\to H$, $t>0, \omega\in\Omega$. Hence the first assertion in (iv) holds.

To prove the strong measurability assertion in (iv), we now highlight the dependence of u on ω. Note first that the map

$$[0,\infty)\times H\times\Omega \ni (t,\psi,\omega)\mapsto u(t,\psi,\omega)\in H$$

is jointly measurable. This is a consequence of the (uniform) continuity of

$$[0,a]\times H \ni (t,\psi)\mapsto u(t,\psi,\omega)\in H, \quad \omega\in\Omega,$$

and the measurability of

$$\Omega\ni\omega\mapsto u(t,\psi,\omega)\in H, \quad (t,\psi)\in \mathbf{R}^+\times H.$$

Secondly, the joint strong measurability of

$$[0,\infty)\times H\times\Omega\ni(t,\psi,\omega)\mapsto Du(t,\psi,\omega)\in L(H)$$

follows from the relation

$$Du(t,\psi,\omega)(\eta) = \lim_{h\to 0}\frac{1}{h}[u(t,\psi+h\eta,\omega)-u(t,\psi,\omega)], \tag{1.4.38}$$

for $(t,\omega)\in\mathbf{R}^+\times\Omega, \psi,\eta\in H$. Finally, note that the integrability estimate in (v) follows from the Lipschitz property of $u(t,\cdot,\omega): H\to H$, $(t,\omega)\in\mathbf{R}^+\times\Omega$. In particular, (1.4.38) and the above Lipschitz property give

$$\|Du(t,\psi,\omega)\|_{L(H)} \le c_5(\omega)$$

for all $(t,\psi,\omega)\in[0,a]\times H\times\Omega$, with $E\log^+ c_5 < \infty$. □

REMARKS.

(i) It is easy to see that the above proof is also valid for the initial boundary value problem with Neumann boundary condition. Note the exact formula of the heat kernel was not needed in the proof. Only estimates such as (1.4.16) and (1.4.17) were actually needed. These kind of estimate holds for Laplacian operator on a bounded domain with smooth boundary and Neumann boundary condition. The generalized solution of (1.4.1) can be defined following Freidlin [Fr]:

$$u(t,\psi)(\xi)$$
$$=\hat{E}\left[\psi(x_t^*)e^{\int_0^t\left(\frac{\Delta(t-s,x_s^*)}{Q(t-s,x_s^*)}+1-|u|^\alpha(t-s,\psi)(x_s^*)\right)ds-\frac{1}{2}\sum_{i=1}^{\infty}\int_0^t\sigma_i^2(x_s^*)ds+\sum_{i=1}^{\infty}\sigma_i(x_s^*)dW^i(t-s)}\right]$$

a.s. Here x_t^* is a diffusion process starting at $\xi \in \mathcal{D}$ with reflection on the boundary $\partial\mathcal{D}$ generated with the operator $\nu\Delta + 2\nu\nabla \log Q(t)\nabla$. One can see that the analysis in the proof of Theorem 1.4.1 carries through for this case as well.

(ii) The dimension restriction is used only to guarantee the Fréchet differentiability of the semiflow in Theorem 1.4.2. This condition is not needed for the existence of the globally Lipschitz flow in Theorem 1.4.1. The conditions in Theorem 1.4.2 are stronger than those in Theorem 1.4.1, and accordingly the result.

(b) *Burgers equation with additive noise*

The stochastic Burgers equation has been studied extensively by many researchers in recent years ([B-C-J], [B-C-F], [D-T-Z], [D-Z.2], [D-D-T], [E-V], [H-L-O-U-Z], [Si], [T-Za], [T-Z]). Here we consider the following stochastic Burgers equation on the interval $[0,1]$,

$$\left.\begin{aligned} du + u\frac{\partial u}{\partial \xi}\,dt =& \nu\Delta u dt + dW(t), \quad t>0,\\ u(0,\psi)(\xi) =& \psi(\xi), \quad \xi\in[0,1],\\ u(t,\psi)(0) =& u(t,\psi)(1)=0, \quad t>0, \end{aligned}\right\} \tag{1.4.39}$$

where the viscosity ν is a positive constant and the initial function $\psi \in L^2([0,1])$. The driving noise W is an $L^2([0,1])$-valued Brownian motion on a complete filtered Wiener space $(\Omega, \bar{\mathcal{F}}, (\mathcal{F})_{t\geq 0}, P)$: That is,

$$W(t) := \sum_{k=1}^{\infty} \sigma_k e_k W^k(t), \quad t \geq 0. \tag{1.4.40}$$

In (1.4.40), e_k, $k \geq 1$, is a complete orthonormal system of eigenfunction of $-\nu\Delta$ with corresponding eigenvalues μ_k, $k \geq 1$. As usual, the W^k, $k \geq 1$, are mutually independent one-dimensional standard Brownian motions on $(\Omega, \bar{\mathcal{F}}, (\mathcal{F})_{t\geq 0}, P)$, and

$$\sum_{k=1}^{\infty} \sigma_k^2 < \infty. \tag{1.4.41}$$

Following ([D-Z.2], pp. 260-265), we will transform the mild solution of the stochastic Burgers equation (1.4.39) to that of (a weak solution of) the random Burgers equation (1.4.42) below: Let $T_t : L^2([0,1]) \to L^2([0,1])$, $t \geq 0$, be the heat semi-group generating the Laplacian $\nu\Delta$ with Dirichlet boundary conditions. Let

$$W^*(t)(\xi) := \int_0^t T_{t-s}dW(s)(\xi), \quad t \geq 0,\ \xi \in [0,1].$$

It is easy to see from (1.4.41) that $W^*(t) \in L^2(\Omega, H_0^1([0,1]))$ for each $t \geq 0$. (Cf. [D-Z.1], Theorem 5.20). Furthermore, $W^* \in L^\infty([0,a], L^2([0,1]))$ a.s. for any finite $a > 0$; that is

$$\|W^*\|_\infty^2 := \sup_{0 \leq t \leq a} \|W^*(t)\|_{L^2([0,1])}^2 < \infty$$

a.s. ([D-Z.1], Theorem 6.10).

Set

$$v(t,\xi) := u(t,\xi) - W^*(t)(\xi), \quad t > 0, \xi \in [0,1].$$

Then $v(t,\xi)$ is a mild solution of the following equation

$$\frac{\partial v}{\partial t} = \nu \Delta v - \frac{1}{2}\frac{\partial}{\partial \xi}(v + W^*(t)(\xi))^2, \quad t > 0, \xi \in [0,1], \tag{1.4.42}$$

in the sense of ([D-Z.2], pp. 260-265).

Viewing equation (1.4.42) as a random Burgers equation, it is not hard to see that, for each initial $\psi \in L^2([0,1])$, it has a unique global solution $v(\cdot,\psi,\omega) \in C(\mathbf{R}^+, L^2([0,1])) \cap L^2([0,1], H^1[0,1])$ for each $\omega \in \Omega$; and for any $a \in \mathbf{R}^+$ and any bounded set $S \subset L^2([0,1])$, the following holds

$$\sup_{\substack{t \in [0,a] \\ \psi \in S}} \|v(t,\psi,\omega)\|_{L^2([0,1])} < \infty \tag{1.4.43}$$

for all $\omega \in \Omega$ (cf. [Ta], Chapter 15, Proposition 1.3; [D-Z.2], pp. 260-265).

A continuous semi-flow for a stochastic Burgers equation with skew-symmetric noise was obtained in [B-C-F]. However, this is not sufficient for our purposes, since we seek to construct random families of *differentiable* stable/unstable manifolds near hyperbolic stationary solutions of (1.4.39). In the following theorem, we establish the existence of a perfect C^1 compacting cocycle for (1.4.39). In Part 2 of this paper, this fact will enable us to use multiplicative ergodic theory techniques in order to prove a local stable/unstable manifold theorem near stationary solutions of the stochastic Burgers equation (1.4.39).

THEOREM 1.4.3. *Consider the stochastic Burgers equation* (1.4.39) *with* $L^2([0,1])$*-valued Brownian* (1.4.40). *Then equation* (1.4.39) *has a mild solution with a* $(\mathcal{B}(\mathbf{R}^+) \otimes \mathcal{B}(L^2([0,1])) \otimes \mathcal{F}, \mathcal{B}(L^2([0,1])))$*-measurable version* $u : \mathbf{R}^+ \times L^2([0,1]) \times \Omega \to L^2([0,1])$ *having the following properties:*

(i) *For each* $\psi \in L^2([0,1])$, $u(\cdot,\psi,\cdot) : \mathbf{R}^+ \times \Omega \to L^2([0,1])$ *is* $(\mathcal{F}_t)_{t \geq 0}$*-adapted.*

(ii) (u,θ) *is a* C^1 *perfect cocycle on* $L^2([0,1])$ *(in the sense of* Definition 1.1.2*).*

(iii) *For each* $(t,\omega) \in (0,\infty) \times \Omega$, *the map* $L^2([0,1]) \ni \psi \mapsto u(t,\psi,\omega) \in L^2([0,1])$ *takes bounded sets into relatively compact sets.*

(iv) *For each* $(t,\psi,\omega) \in (0,\infty) \times L^2([0,1]) \times \Omega$, *the Fréchet derivative* $Du(t,\psi,\omega) \in L(L^2([0,1]))$ *is compact. Furthermore, the map*

$$[0,\infty) \times L^2([0,1]) \times \Omega \ni (t,\psi,\omega) \mapsto Du(t,\psi,\omega) \in L(L^2([0,1]))$$

is strongly measurable.

(v) *For any positive reals* a, ρ,

$$E \log^+ \sup_{\substack{0 \leq t \leq a \\ \|\psi\|_{L^2([0,1])} \leq \rho}} \left\{ \|u(t,\psi,\cdot)\|_{L^2([0,1])} + \|Du(t,\psi,\cdot)\|_{L(L^2([0,1]))} \right\} < \infty.$$

PROOF. For simplicity of notation, we will assume throughout this proof that $\nu = \frac{1}{2}$.

Assertion (i) follows easily from the global existence of solutions to (1.4.42).

To prove (ii), consider Burgers equation (1.4.42). Denote by $p(t,\xi,y)$ the heat kernel for the Laplacian $\nu\Delta$ on $[0,1]$ with Dirichlet boundary conditions. Recall that there are positive constants c_1, c_2 such that

$$\left|\frac{\partial p(t,\xi,y)}{\partial y}\right| \leq \frac{c_1}{t} e^{-\frac{(\xi-y)^2}{2c_2 t}} \tag{1.4.44}$$

for all $t > 0$, $\xi, y \in [0,1]$ (c.f. [L-S-U], p. 413). Then pick a positive constant c_3 such that

$$\int_{-\infty}^{\infty} \frac{c_3}{\sqrt{t}} e^{-\frac{y^2}{2c_2 t}}\, dy \leq 1$$

for all $t > 0$.

Using (1.4.42), variation of parameters, and integration by parts, we get

$$\begin{aligned}
v(t,\psi)(\xi) =& T_t\psi(\xi) - \frac{1}{2}\int_0^t T_{t-s}\nabla v^2(s,\psi)(\xi)\, ds \\
&+ \int_0^t T_{t-s}(-\nabla(W^*(s)v(s,\psi)) - W^*(s)\nabla W^*(s))(\xi)\, ds \\
=& \int_0^1 p(t,\xi,y)\psi(y)dy - \frac{1}{2}\int_0^t\int_0^1 p(t-s,\xi,y)\nabla v^2(s,\psi)(y)\, dy\, ds \\
&+ \int_0^t\int_0^1 p(t-s,\xi,y)(-\nabla(W^*(s)v(s,\psi)) - W^*(s)\nabla W^*(s))(y)\, dy\, ds \\
=& \int_0^1 p(t,\xi,y)\psi(y)dy + \frac{1}{2}\int_0^t\int_0^1 \nabla p(t-s,\xi,y)v^2(s,\psi)(y)\, dy\, ds \\
&+ \int_0^t\int_0^1 \nabla p(t-s,\xi,y)(W^*(s)v(s,\psi) + \frac{1}{2}W^*(s)^2)(y)\, dy\, ds
\end{aligned}$$

for all $t \geq 0$, $\xi \in [0,1]$. Thus

$$\begin{aligned}
v(t,\psi+hg)(\xi) - v(t,\psi)(\xi) &= h\int_0^1 p(t,\xi,y)g(y)dy \\
&+ \frac{1}{2}\int_0^t\int_0^1 \nabla p(t-s,\xi,y)(v^2(s,\psi+hg)(y) - v^2(s,\psi)(y))\, dy\, ds \\
&+ \int_0^t\int_0^1 \nabla p(t-s,\xi,y)(W^*(s)(v(s,\psi+hg) - v(s,\psi))(y)\, dy\, ds,
\end{aligned}$$

for all $t \geq 0$, $\xi \in [0,1]$, a.s.. Squaring both sides of the above equality, integrating with respect $\xi \in [0,1]$, and using the heat kernel estimate (1.4.44), we obtain

$$\begin{aligned}
&\|v(t,\psi+hg) - v(t,\psi)\|^2_{L^2([0,1])} \leq 3h^2\|g\|^2_{L^2([0,1])} \\
&+ \frac{3}{4}\int_0^1\left(\int_0^t\int_0^1 \nabla p(t-s,\xi,y)(v^2(s,\psi+hg)(y) - v^2(s,\psi)(y))dyds\right)^2 d\xi \\
&+ 3\int_0^1\left(\int_0^t\int_0^1 \nabla p(t-s,\xi,y)W^*(s)(y)(v(s,\psi+hg)(y) - v(s,\psi)(y))dyds\right)^2 d\xi
\end{aligned}$$

$$
\begin{aligned}
\leq & 3h^2\|g\|^2_{L^2([0,1])} \\
& + \frac{3}{4}\int_0^1 \Big(\int_0^t \frac{1}{\sqrt{t-s}} \int_0^1 \frac{c_1}{\sqrt{t-s}} \mathrm{e}^{-\frac{(\xi-y)^2}{2c_2(t-s)}} (v^2(s,\psi+hg)(y) \\
& \qquad\qquad - v^2(s,\psi)(y))dyds\Big)^2 d\xi \\
& + 3\int_0^1 \Big(\int_0^t \frac{1}{\sqrt{t-s}} \int_0^1 \frac{c_1}{\sqrt{t-s}} \mathrm{e}^{-\frac{(\xi-y)^2}{2c_2(t-s)}} W^*(s)(y) \times \\
& \qquad\qquad \times (v(s,\psi+hg)(y) - v(s,\psi)(y))\, dyds\Big)^2 d\xi,
\end{aligned}
$$

for all $t > 0$, a.s..

From now on, we will denote by the same letter C all generic positive and (possibly) random constants that may change from line to line.

Using the Cauchy-Schwartz inequality in the above inequality, the heat kernel estimate (1.4.44) and Fubini's theorem, we obtain

$$
\begin{aligned}
\|v(t,&\psi+hg) - v(t,\psi)\|^2_{L^2([0,1])} \leq 3h^2\|g\|^2_{L^2([0,1])} + \frac{3}{4}\int_0^1 \int_0^t \frac{1}{(t-s)^{\frac{3}{4}}} ds \\
& \int_0^t \frac{1}{(t-s)^{\frac{1}{4}}} \Big(\int_0^1 \frac{c_1}{\sqrt{t-s}} \mathrm{e}^{-\frac{(\xi-y)^2}{2c_2(t-s)}} (v^2(s,\psi+hg)(y) - v^2(s,\psi)(y))dy\Big)^2 ds\, d\xi \\
& + 3\int_0^1 \int_0^t \frac{1}{(t-s)^{\frac{3}{4}}} ds \int_0^t \frac{1}{(t-s)^{\frac{1}{4}}} \\
& \Big(\int_0^1 \frac{c_1}{\sqrt{t-s}} \mathrm{e}^{-\frac{(\xi-y)^2}{2c_2(t-s)}} W^*(s)(y)(v(s,\psi+hg) - v(s,\psi))(y)dy\Big)^2 ds\, d\xi \\
\leq & 3h^2\|g\|^2_{L^2([0,1])} + Ct^{\frac{1}{4}} \int_0^1 \int_0^t \frac{1}{(t-s)^{\frac{1}{4}}} \\
& \quad \times \int_0^1 \frac{c_1}{\sqrt{t-s}} \mathrm{e}^{-\frac{(\xi-y)^2}{2c_2(t-s)}} (v^2(s,\psi+hg)(y) + v^2(s,\psi)(y))dy \\
& \quad \times \int_0^1 \frac{c_1}{\sqrt{t-s}} \mathrm{e}^{-\frac{(\xi-y)^2}{2c_2(t-s)}} (v(s,\psi+hg)(y) - v(s,\psi)(y))^2 dy\, ds\, d\xi \\
& + Ct^{\frac{1}{4}} \int_0^1 \int_0^t \frac{1}{(t-s)^{\frac{1}{4}}} \int_0^1 \frac{c_1}{\sqrt{t-s}} \mathrm{e}^{-\frac{(\xi-y)^2}{2c_2(t-s)}} W^*(s)(y)^2 dy \\
& \quad \times \int_0^1 \frac{c_1}{\sqrt{t-s}} \mathrm{e}^{-\frac{(\xi-y)^2}{2c_2(t-s)}} (v(s,\psi+hg) - v(s,\psi))^2(y) dy\, ds\, d\xi \\
\leq & 3h^2\|g\|^2_{L^2([0,1])} + Ct^{\frac{1}{4}} \int_0^t \frac{1}{(t-s)^{\frac{3}{4}}} \int_0^1 (v^2(s,\psi+hg)(y) + v^2(s,\psi)(y))dy \\
& \quad \times \int_0^1 \int_0^1 \frac{c_1}{\sqrt{t-s}} \mathrm{e}^{-\frac{(\xi-y)^2}{2c_2(t-s)}} d\xi (v(s,\psi+hg)(y) - v(s,\psi)(y))^2 dy\, ds \\
& + Ct^{\frac{1}{4}} \int_0^t \frac{1}{(t-s)^{\frac{3}{4}}} \int_0^1 W^*(s)(y)^2 dy
\end{aligned}
$$

$$\times \int_0^1 \int_0^1 \frac{c_1}{\sqrt{t-s}} e^{-\frac{(\xi-y)^2}{2c_2(t-s)}} d\xi (v(s,\psi+hg)-v(s,\psi))^2(y) dy\, ds$$

$$\leq 3h^2\|g\|^2_{L^2([0,1])} + Ct^{\frac{1}{4}} \int_0^t \frac{1}{(t-s)^{\frac{3}{4}}} \|v(s,\psi+hg)-v(s,\psi)\|^2_{L^2([0,1])}\, ds$$

$$+ C\|W^*\|^2_\infty t^{\frac{1}{4}} \int_0^t \frac{1}{(t-s)^{\frac{3}{4}}} \|v(s,\psi+hg)-v(s,\psi)\|^2_{L^2([0,1])}\, ds,$$

$$\leq 3h^2\|g\|^2_{L^2([0,1])} + Ct^{\frac{1}{4}} \int_0^t \frac{1}{(t-s)^{\frac{3}{4}}} \|v(s,\psi+hg)-v(s,\psi)\|^2_{L^2([0,1])}\, ds,$$

for all $t>0$, a.s.. Iterating the above computation, we get

$$\|v(t,\psi+hg)-v(t,\psi)\|^2_{L^2([0,1])}$$

$$\leq 3h^2\|g\|^2_{L^2([0,1])} + Ct^{\frac{1}{4}}\, h^2\|g\|^2_{L^2([0,1])} \int_0^t \frac{1}{(t-s)^{\frac{3}{4}}}\, ds$$

$$+ Ct^{\frac{1}{4}} \int_0^t \int_0^s \frac{s^{\frac{1}{4}}}{(t-s)^{\frac{3}{4}}(s-r)^{\frac{3}{4}}} \|v(r,\psi+hg)-v(r,\psi)\|^2_{L^2([0,1])}\, dr\, ds$$

for all $t>0$, a.s.. Consider now the elementary estimate

$$\int_r^t \frac{s^\alpha}{(t-s)^\beta (s-r)^\gamma} ds = \int_0^{t-r} \frac{(s+r)^\alpha}{(t-r-s)^\beta s^\gamma} ds \leq \frac{C_1}{(t-r)^{\beta+\gamma-1}}, \quad t \geq r > 0,$$

which holds for any $\alpha \geq 0$, $0 \leq \beta < 1$, $0 \leq \gamma < 1$, and where $C_1 > 0$ is a positive (deterministic) constant. Using the above estimate together with Fubini's theorem, gives

$$\|v(t,\psi+hg)-v(t,\psi)\|^2_{L^2([0,1])}$$

$$\leq 3h^2\|g\|^2_{L^2([0,1])} + Ct^{\frac{1}{2}} h^2\|g\|^2_{L^2([0,1])}$$

$$+ Ct^{\frac{1}{4}} \int_0^t \|v(r,\psi+hg)-v(r,\psi)\|^2_{L^2([0,1])} \int_r^t \frac{s^{\frac{1}{4}}}{(t-s)^{\frac{3}{4}}(s-r)^{\frac{3}{4}}}\, ds\, dr$$

$$\leq 3h^2\|g\|^2_{L^2([0,1])} + Ct^{\frac{1}{2}} h^2\|g\|^2_{L^2([0,1])}$$

$$+ Ct^{\frac{1}{4}} \int_0^t \frac{1}{(t-r)^{1/2}} \|v(r,\psi+hg)-v(r,\psi)\|^2_{L^2([0,1])} dr \tag{1.4.45}$$

a.s. for all $t \in (0,a]$, $a \in \mathbf{R}^+$. Iterating the above process once more and applying Gronwall's lemma, we obtain

$$\sup_{\substack{0\leq t\leq a,\, g\in L^2([0,1])\\ \|g\|_{L^2}\leq 1}} \|v(t,\psi+hg)-v(t,\psi)\|^2_{L^2([0,1])} \leq Mh^2, \tag{1.4.46}$$

a.s. for any $a \in \mathbf{R}^+$, where M is a positive random constant depending on a.

For fixed $\psi, g \in L^2([0,1])$, define $G := G(t,\psi)(g)(\xi)$, $t>0, \xi \in [0,1]$, to be the weak solution of the "linearized" Burgers equation

$$\frac{\partial G}{\partial t} + \frac{\partial (v(t,\psi)G)}{\partial \xi} = \frac{1}{2}\Delta G - \frac{\partial (W^* G)}{\partial \xi}, \quad G(0,\psi)(g) = g \in L^2([0,1]).$$

Set

$$\mu_t(\psi, g) := v(t, \psi + hg) - v(t, \psi) - hG(t, \psi)(g), \quad |h| < 1, t \geq 0.$$

Then it is easy to see that

$$\begin{aligned}\mu_t(\psi, g)(\xi) = &- \int_0^t \int_0^1 p(t-s, \xi, y)(\frac{1}{2}\nabla(v(s, \psi + hg)(y) - v(s, \psi)(y))^2 \\ &+ \nabla(v(s, \psi)(y)\mu_s(\psi, g)(y)) + \nabla(W^*(s)(y)\mu_s(\psi, g)(y)))\, dy\, ds \\ = &\int_0^t \int_0^1 \nabla p(t-s, \xi, y)(\frac{1}{2}(v(s, \psi + hg)(y) - v(s, \psi)(y))^2 \\ &+ v(s, \psi)(y)\mu_s(\psi, g)(y) + W^*(s)(y)\mu_s(\psi, g)(y))\, dy\, ds\end{aligned}$$

a.s. for all $t > 0$. So using the Cauchy-Schwartz inequality and (1.4.46), we obtain

$$\begin{aligned}&\|\mu_t(\psi, g)\|^2_{L^2([0,1])} \\ \leq &\frac{3}{4}\int_0^1 \left(\int_0^t \int_0^1 \nabla p(t-s, \xi, y)(v(s, \psi + hg)(y) - v(s, \psi)(y))^2\, dy\, ds\right)^2 d\xi \\ &+ 3\int_0^1 \left(\int_0^t \int_0^1 \nabla p(t-s, \xi, y)(v(s, \psi)(y)\mu_s(\psi, g)(y)\, dy\, ds\right)^2 d\xi \\ &+ 3\int_0^1 \left(\int_0^t \int_0^1 \nabla p(t-s, \xi, y)(W^*(s)(y)\mu_s(\psi, g)(y))\, dy\, ds\right)^2 d\xi \\ \leq &\frac{3}{4}\int_0^1 \int_0^t \frac{1}{(t-s)^{\frac{3}{4}}}\, ds \int_0^t \frac{1}{(t-s)^{\frac{1}{4}}} \left(\int_0^1 \frac{c_1}{\sqrt{t-s}} e^{-\frac{(y-\xi)^2}{c_2(t-s)}}\right. \\ &\left.(v(s, \psi + hg)(y) - v(s, \psi)(y))^2 dy\right)^2 ds\, d\xi \\ &+ 3\int_0^1 \int_0^t \frac{1}{(t-s)^{\frac{3}{4}}} ds \int_0^t \frac{1}{(t-s)^{\frac{1}{4}}} \left(\int_0^1 \frac{c_1}{\sqrt{t-s}} e^{-\frac{(y-\xi)^2}{c_2(t-s)}}\right. \\ &\left.(v(s, \psi)(y)\mu_s(\psi, g)(y) dy\right)^2 ds d\xi \\ &+ 3\int_0^1 \int_0^t \frac{1}{(t-s)^{\frac{3}{4}}}\, ds \int_0^t \frac{1}{(t-s)^{\frac{1}{4}}} \left(\int_0^1 \frac{c_1}{\sqrt{t-s}} e^{-\frac{(y-\xi)^2}{c_2(t-s)}}\right. \\ &\left.(W^*(s)(y)\mu_s(\psi, g)(y)) dy\right)^2 ds\, d\xi.\end{aligned}$$

a.s. for all $t > 0$. Thus

$$\begin{aligned}&\|\mu_t(\psi, g)\|^2_{L^2([0,1])} \\ \leq &3c_1 t^{\frac{1}{4}} \int_0^t \frac{1}{(t-s)^{\frac{3}{4}}} \left(\int_0^1 (v(s, \psi + hg)(y) - v(s, \psi)(y))^2 dy\right)^2 ds \\ &+ 12c_1 t^{\frac{1}{4}} \int_0^t \frac{1}{(t-s)^{\frac{3}{4}}} \int_0^1 v^2(s, \psi)(y) dy \int_0^1 \mu_s^2(\psi, g)(y)\, dy\, ds \\ &+ Ct^{\frac{1}{4}} \|W^*\|^2_\infty \int_0^t \frac{1}{(t-s)^{\frac{3}{4}}} \int_0^1 \mu_s^2(\psi, g)(y)\, dy\, ds\end{aligned}$$

$$\leq Ch^4 + Ct^{\frac{1}{4}} \int_0^t \frac{1}{(t-s)^{\frac{3}{4}}} \|\mu_s(\psi, g\|^2_{L^2([0,1])}\, ds, \quad 0 \leq t \leq a,$$

a.s.. Using the previous iteration argument followed by Gronwall's lemma, we obtain the following estimate

$$\sup_{\substack{g \in L^2([0,1]), \|g\|_{L^2} \leq 1 \\ 0 \leq t \leq a}} \|\mu_t(\psi, g)\|^2_{L^2([0,1])} \leq M_1 h^4, \quad |h| < 1$$

a.s. for some positive random constant $M_1 = M_1(\omega, a)$. This implies that $\dfrac{\mu_t(\psi, g)}{h}$ converges a.s. to 0 as $h \to 0$ in $L^2([0,1])$, uniformly in $(t,g) \in [0,a] \times \{g \in L^2([0,1]) : \|g\|_{L^2} \leq 1\}$. Therefore, $\dfrac{v(t, \psi + hg) - v(t, \psi)}{h} \to G_t(\psi, g)$ a.s. as $h \to 0$, uniformly for $g \in \{g; \|g\|_{L^2([0,1])} \leq 1\}$. Hence, v is Fréchet differentiable at $\psi \in L^2([0,1])$, with Fréchet derivative $Dv(t,\psi) : L^2([0,1]) \to L^2([0,1])$ satisfying the $L(L^2([0,1]))$-valued linear equation

$$\begin{aligned} Dv(t,\psi) =& T_t - \int_0^t T_{t-s} \left(\frac{\partial v(s,\psi)}{\partial \xi} Dv(s,\psi) + v(s,\psi) \frac{\partial Dv(s,\psi)}{\partial \xi} \right) ds \\ & - \int_0^t T_{t-s} \left(Dv(s,\psi) \frac{\partial W^*(s)}{\partial \xi} + W^*(s) \frac{\partial Dv(s,\psi)}{\partial \xi} \right) ds \end{aligned} \tag{1.4.47}$$

a.s. for $(t, \psi) \in \mathbf{R}^+ \times L^2([0,1])$.

In order to complete the proof of assertion (ii) of the theorem, it remains to prove that (u, θ) is a perfect cocycle in $L^2([0,1])$. It is easy to see from (1.4.39) that

$$u(t,\psi)(\omega) = T_t \psi - \int_0^t T_{t-s} u(s,\psi,\omega) \nabla u(s,\psi,\omega)\, ds + \left[\int_0^t T_{t-s} dW(s) \right](\omega)$$

for $t > 0, \omega \in \Omega, \psi \in L^2([0,1])$. We need to prove that

$$u(t, u(t_1, \psi, \omega), \theta(t_1, \omega)) = u(t + t_1, \psi, \omega). \tag{1.4.48}$$

for $t, t_1 \geq 0, \omega \in \Omega, \psi \in L^2([0,1])$. To see this, fix $t_1 \geq 0, \omega \in \Omega, \psi \in L^2([0,1])$, and denote

$$Y(t) := u(t, u(t_1, \psi, \omega), \theta(t_1, \omega)), \quad Z(t) := u(t + t_1, \psi, \omega), \quad t > 0.$$

Then

$$\begin{aligned} Y(t) =& T_t u(t_1, \psi, \omega) - \int_0^t T_{t-s} u(s, u(t_1,\psi,\omega), \theta(t_1,\omega)) \frac{\partial u(s, u(t_1,\psi,\omega), \theta(t_1,\omega))}{\partial y} ds \\ & + \left[\int_{t_1}^{t+t_1} T_{t+t_1-s} dW(s) \right](\omega) \\ =& T_{t+t_1} \psi - \int_0^{t_1} T_{t+t_1-s} u(s,\psi,\omega) \frac{\partial u(s,\psi,\omega)}{\partial y} ds \\ & - \int_{t_1}^{t+t_1} T_{t+t_1-s} Y(s-t_1) \frac{\partial Y(s-t_1)}{\partial y} ds + \left[\int_0^{t+t_1} T_{t+t_1-s} dW(s) \right](\omega), \; t > 0. \end{aligned}$$

Also,

$$\begin{aligned}Z(t) =&T_{t+t_1}\psi - \int_0^{t_1} T_{t+t_1-s}u(s,\psi,\omega)\frac{\partial u(s,\psi,\omega)}{\partial y}ds \\ &- \int_{t_1}^{t+t_1} T_{t+t_1-s}Z(s-t_1)\frac{\partial Z(s-t_1)}{\partial y}ds + \left[\int_0^{t+t_1} T_{t+t_1-s}dW(s)\right](\omega),\ t>0.\end{aligned}$$

Therefore,

$$\begin{aligned}&Y(t)(\xi) - Z(t)(\xi) \\ &= -\frac{1}{2}\int_{t_1}^{t+t_1} T_{t+t_1-s}\left(\frac{\partial Y^2(s-t_1)}{\partial y} - \frac{\partial Z^2(s-t_1)}{\partial y}\right)(\xi)ds \\ &=\frac{1}{2}\int_{t_1}^{t+t_1} ds\int_0^1 \frac{\partial p(t+t_1-s,\xi,y)}{\partial y}[Y^2(s-t_1)(y) - Z^2(s-t_1)(y)]\,dy \\ &=\frac{1}{2}\int_0^t ds\int_0^1 \frac{\partial p(t-s,\xi,y)}{\partial y}[Y^2(s)(y) - Z^2(s)(y)]\,dy,\end{aligned}$$

a.s. for $t>0, \xi\in\mathcal{D}$. Using Cauchy-Schwartz inequality and (1.4.43), we have

$$\begin{aligned}&\|Y(t)-Z(t)\|^2_{L^2([0,1])} = \int_0^1 |Y(t)(\xi)-Z(t)(\xi)|^2\,d\xi \\ \leq& C\int_0^1 d\xi\left(\frac{1}{2}\int_0^t ds\int_0^1 \frac{1}{\sqrt{t-s}}\frac{c_1}{\sqrt{t-s}}e^{-\frac{(\xi-y)^2}{2c_2(t-s)}}|Y^2(s)(y)-Z^2(s)(y)|dy\right)^2 \\ \leq& C\int_0^1\int_0^t \frac{1}{(t-s)^{\frac{3}{4}}}ds\int_0^t\frac{1}{(t-s)^{\frac{1}{4}}} \\ &\times\left(\int_0^1 \frac{c_1}{\sqrt{t-s}}e^{-\frac{(\xi-y)^2}{2c_2(t-s)}}|Y^2(s)(y)-Z^2(s)(y)|dy\right)^2 dsd\xi \\ \leq& Ct^{\frac{1}{4}}\int_0^1\int_0^t \frac{1}{(t-s)^{\frac{1}{4}}}\int_0^1 \frac{c_1}{\sqrt{t-s}}e^{-\frac{(\xi-y)^2}{2c_2(t-s)}}|Y(s)(y)-Z(s)(y)|^2\,dy \\ &\times\int_0^1 \frac{c_1}{\sqrt{t-s}}e^{-\frac{(\xi-y)^2}{2c_2(t-s)}}[Y(s)(y)+Z(s)(y)]^2\,dy\,ds\,d\xi \\ \leq& Ct^{\frac{1}{4}}\sup_{0\leq s\leq T}[\|Y(s)+Z(s)\|^2_{L^2([0,1])}]\int_0^t\frac{1}{(t-s)^{\frac{3}{4}}}\|Y(s)-Z(s)\|^2_{L^2([0,1])}ds \\ \leq& Ct^{\frac{1}{4}}\int_0^t \frac{1}{(t-s)^{\frac{3}{4}}}\|Y(s)-Z(s)\|^2_{L^2([0,1])}ds,\end{aligned} \tag{1.4.49}$$

for all $t\in(0,a]$, a.s.. Note that in the above computation, we have also used the fact that

$$\sup_{0\leq s\leq a}[\|Y(s)+Z(s)\|^2_{L^2([0,1])}] < \infty, \quad a\in\mathbf{R}^+,$$

a.s.. As in the proof of (1.4.45) it follows from (1.4.49) and Gronwall's lemma that $Y(t) = Z(t)$ a.s. for all $t\geq 0$. This completes the proof of assertion (ii) of the theorem.

We next prove assertion (iii). It is not hard to see that

(1.4.50)
$$\begin{aligned}\|v(t,\psi_m,\omega)-v(t,\psi_n,\omega)\|^2_{L^2([0,1])} &\le C\|T_t\psi_m - T_t\psi_n\|^2_{L^2([0,1])} \\ &+ Ct^{\frac{1}{4}}\int_0^t \frac{1}{(t-s)^{\frac{3}{4}}}\|v(s,\psi_m,\omega)-v(s,\psi_n,\omega)\|^2_{L^2([0,1])}\,ds\end{aligned}$$

a.s. for all $t \in [0,a]$. Now using (1.4.50) and the same argument as in the proof of compactness of $Dv(t,\psi,\omega)$ in Theorem 1.4.2, one can show that, for each $t > 0, \omega \in \Omega$, the map $v(t,\cdot,\omega) : L^2([0,1]) \to L^2([0,1])$ takes bounded sets into relatively compact sets. The only difference is that we have to iterate (1.4.50) once before we can use Gronwall's lemma. Details of the proof are omitted. $\square$

Part 2

Existence of Stable and Unstable Manifolds

2.1. Hyperbolicity of a Stationary Trajectory

In Part 1, we established the existence of perfect differentiable cocycles generated by mild solutions of a large class of semilinear stochastic evolution equations (see's) and stochastic partial differential equations (spde's). In this part, we continue the analysis in Part 1. More specifically, we highlight the concept of a *stationary point* for the see or spde as an invariant random vector under the cocycle. Our main objective is to characterize the pathwise local structure of solutions of semilinear see's and spde's near stationary solutions. We define the concept of *hyperbolicity* for a stationary solution of an see/spde. Hyperbolicity is characterized by the non-vanishing of the Lyapunov spectrum of the linearized cocycle. The hyperbolic structure of the stochastic semiflow leads to *local stable manifold theorems* (Theorems 2.4.1-2.4.4) for semilinear see's and spde's. For a hyperbolic stationary solution of the see/spde, this gives smooth stable and unstable manifolds in a neighborhood of the stationary solution. The stable and unstable manifolds are stationary, live in a stationary tubular neighborhood of the stationary solution and are asymptotically invariant under the stochastic semiflow of the see/spde. Furthermore, the local stable and unstable manifolds intersect transversally at the stationary point, and the unstable manifolds have fixed finite dimension. Due to their forward asymptotic dependence on the future of the stochastic semiflow, the stable and unstable manifolds are in general anticipating in nature. In particular, the tangent spaces to the stable and unstable manifolds (at the stationary point) are constructed using the (anticipating) eigenspaces of the Oseledec-Ruelle operator for the hyperbolic linearized cocycle.

The proof of the stable manifold theorem (Theorem 2.2.1) uses discrete infinite-dimensional multiplicative ergodic theory techniques ([Ru.1], [Ru.2], Theorems 5.1, 6.1). The extension to continuous-time is done via perfection techniques and interpolation between discrete times (cf. [Mo.3], [M-S.2], [M-S.1]). Henceforth, we will assume that the reader is familiar with the results and the techniques in Ruelle's articles [Ru.1] and [Ru.2].

We recall below the definition of a cocycle in Hilbert space.

Let $(\Omega, \mathcal{F}, P)$ be a probability space. Suppose $\theta : \mathbf{R}\times\Omega \to \Omega$ is a $(\mathcal{B}(\mathbf{R})\otimes\mathcal{F}, \mathcal{F})$-measurable group of P-preserving ergodic transformations on $(\Omega, \mathcal{F}, P)$. Denote by $\bar{\mathcal{F}}$ the P-completion of $\mathcal{F}$, and suppose $(\Omega, \bar{\mathcal{F}}, (\mathcal{F})_{t\geq 0}, P)$ is a complete filtered probability space.

Let H be a real separable Hilbert space with norm $|\cdot|$ and Borel σ-algebra $\mathcal{B}(H)$.

Take k to be any non-negative integer and $\epsilon \in (0,1]$. Recall that a $C^{k,\epsilon}$ *perfect cocycle* (U,θ) on H is a $(\mathcal{B}(\mathbf{R}^+) \otimes \mathcal{B}(H) \otimes \mathcal{F}, \mathcal{B}(H))$- measurable random field $U : \mathbf{R}^+ \times H \times \Omega \to H$ with the following properties:

(i) For each $\omega \in \Omega$, the map $\mathbf{R}^+\times H \ni (t,x) \mapsto U(t,x,\omega) \in H$ is jointly continuous; for fixed $(t,\omega) \in \mathbf{R}^+\times\Omega$, the map $H \ni x \mapsto U(t,x,\omega) \in H$ is $C^{k,\epsilon}$ ($D^kU(t,x,\omega)$ is C^ϵ in x on bounded subsets of H).

(ii) $U(t_1+t_2,\cdot,\omega) = U(t_2,\cdot,\theta(t_1,\omega)) \circ U(t_1,\cdot,\omega)$ for all $t_1, t_2 \in \mathbf{R}^+$, all $\omega \in \Omega$.

(iii) $U(0,x,\omega) = x$ for all $x \in H, \omega \in \Omega$.

We next describe the concept of a *stationary point* for a cocycle (U,θ). Stationary points play the role of stochastic equilibria for the stochastic dynamical system.

DEFINITION 2.1.1. An $\mathcal{F}$-measurable random variable $Y : \Omega \to H$ is said be a *stationary random point* for the cocycle (U, θ) if it satisfies the following identity:

$$U(t, Y(\omega), \omega) = Y(\theta(t, \omega)) \tag{2.1.1}$$

for all $(t, \omega) \in \mathbf{R}^+ \times \Omega$.

The reader may note that the above definition is an infinite-dimensional analogue of a corresponding concept of invariance that was used by one of the authors in joint work with M. Scheutzow to give a proof of the stable manifold theorem for stochastic ordinary differential equations (Definition 3.1, [M-S.2]). Definition 2.1.1 essentially gives a useful realization of the idea of an invariant measure for a stochastic dynamical system generated by an spde or a see. Such a realization allows us to analyze the local *almost sure* stability properties of the stochastic semiflow in the neighborhood of the stationary point. The existence (and uniqueness/ergodicity) of a stationary random point for various classes of spde's and see's has been studied by many researchers. In this article, we move beyond the issue of existence of stationary solutions, and apply our stable/unstable manifold theorem to examine further the almost sure asymptotic structure of the stochastic flow generated by several well-known classes of see's and spde's. In particular, we establish the existence of local stable and unstable manifolds near their stationary points.

Note that, in general, $Y(\theta(t, \omega))$ is not an adapted process because the stationary point Y may depend on the full Brownian path that drives the spde: See Proposition 2.4.1. Thus, one does not expect that $Y(\theta(t, \omega))$ would solve the underlying *Itô-type* see or spde. However, it has been established in joint work by one of the authors with M. Scheutzow that, for sode's, such a stationary trajectory does indeed satisfy the corresponding *Stratonovich* version of the sde ([M-S.2], Theorem A.2). In our present context, we conjecture that an analogous result also holds for the see's and spde's treated in this article.

The main objective of this section is to define the concept of *hyperbolicity* for a stationary point Y of the cocycle (U, θ).

First, we linearize the $C^{k,\epsilon}$ cocycle (U, θ) along a stationary random point Y. By taking Fréchet derivatives at the stationary point $Y(\omega)$ on each side of the cocycle identity (ii) above, using the chain rule and the definition of Y, we immediately see that $(DU(t, Y(\omega), \omega), \theta(t, \omega))$ is an $L(H)$-valued perfect cocycle. Secondly, we appeal to the following classical result which goes back to Oseledec in the finite-dimensional case ([O]), and to D. Ruelle in infinite dimensions ([Ru.2]).

THEOREM 2.1.1. *(Oseledec-Ruelle)*

Let $T : \mathbf{R}^+ \times \Omega \to L(H)$ be strongly measurable, such that (T, θ) is an $L(H)$-valued cocycle, with each $T(t, \omega)$ compact. Suppose that

$$E \sup_{0 \le t \le 1} \log^+ \|T(t, \cdot)\|_{L(H)} + E \sup_{0 \le t \le 1} \log^+ \|T(1 - t, \theta(t, \cdot))\|_{L(H)} < \infty.$$

Then there is a sure event $\Omega_0 \in \mathcal{F}$ such that $\theta(t, \cdot)(\Omega_0) \subseteq \Omega_0$ for all $t \in \mathbf{R}^+$, and for each $\omega \in \Omega_0$, the limit

$$\Lambda(\omega) := \lim_{t \to \infty} [T(t, \omega)^* \circ T(t, \omega)]^{1/(2t)}$$

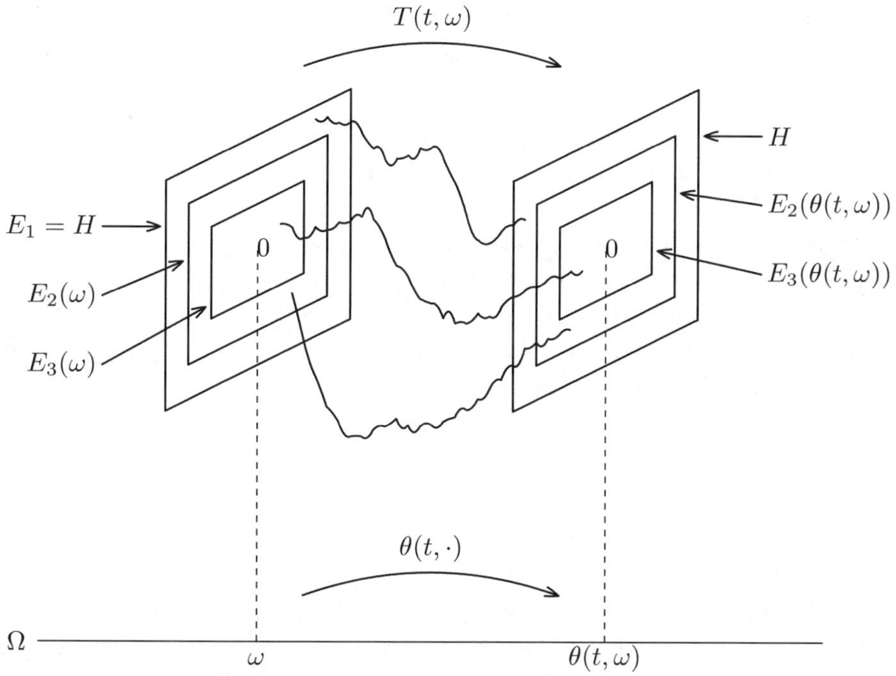

FIGURE 2. The Spectral Theorem.

exists in the uniform operator norm. Each linear operator $\Lambda(\omega)$ is compact, non-negative and self-adjoint with a discrete spectrum

$$e^{\lambda_1} > e^{\lambda_2} > e^{\lambda_3} > \cdots$$

where the λ_i's are distinct and non-random. Each positive eigenvalue $e^{\lambda_i}(> 0)$ has a fixed finite non-random multiplicity m_i and a corresponding eigenspace $F_i(\omega)$, with $m_i := dim F_i(\omega)$. Set $i = \infty$ when $\lambda_i = -\infty$. Define

$$E_1(\omega) := H, \quad E_i(\omega) := \left[\oplus_{j=1}^{i-1} F_j(\omega)\right]^{\perp}, \ i > 1, \ E_\infty := ker \Lambda(\omega).$$

Then

$$E_\infty \subset \cdots \subset \cdots \subset E_{i+1}(\omega) \subset E_i(\omega) \cdots \subset E_2(\omega) \subset E_1(\omega) = H,$$

$$\lim_{t\to\infty} \frac{1}{t} \log |T(t,\omega)x| = \begin{cases} \lambda_i & \text{if} \quad x \in E_i(\omega)\backslash E_{i+1}(\omega), \\ -\infty & \text{if} \quad x \in E_\infty(\omega), \end{cases}$$

and

$$T(t,\omega)(E_i(\omega)) \subseteq E_i(\theta(t,\omega))$$

for all $t \geq 0, \ i \geq 1$.

Fig. 2 illustrates the Oseledec-Ruelle theorem.

PROOF OF THEOREM 2.1.1. The proof is based on a discrete version of Oseledec's multiplicative ergodic theorem and the perfect ergodic theorem ([Ru.1], I.H.E.S Publications, 1979, pp. 303-304; cf. [O], [Mo.3], Lemma 5. See also Lemma 2.3.1 (ii) of this article). Details of the extension to continuous time are given in [Mo.3] within the context of linear stochastic functional differential equations. The arguments in [Mo.3] extend directly to general linear cocycles in Hilbert space. cf. [F-S.1]. □

DEFINITION 2.1.2. The sequence $\{\cdots < \lambda_{i+1} < \lambda_i < \cdots < \lambda_2 < \lambda_1\}$ in the Oseledec-Ruelle theorem (Theorem 2.1.1) is called the *Lyapunov spectrum* of the linear cocycle (T, θ).

Hyperbolicity of a stationary point $Y : \Omega \to H$ of the non-linear cocycle (U, θ) may now be defined in terms of a spectral gap in the Lyapunov spectrum of the linearized cocycle $(DU(t, Y(\omega), \omega), \theta(t, \omega))$.

DEFINITION 2.1.3. Let (U, θ) be a $C^{k,\epsilon}$ $(k \geq 1, \epsilon \in (0, 1])$ perfect cocycle on a separable Hilbert space H such that $U(t, \cdot, \omega) : H \to H$ takes bounded sets into relatively compact sets for each $(t, \omega) \in (0, \infty) \times \Omega$. A stationary point $Y(\omega)$ of the cocycle (U, θ) is *hyperbolic* if

(a) For any $a \in (0, \infty)$,

$$\int_\Omega \log^+ \sup_{0 \leq t_1, t_2 \leq a} \|DU(t_2, Y(\theta(t_1, \omega)), \theta(t_1, \omega))\|_{L(H)} \, dP(\omega) < \infty.$$

(b) The linearized cocycle $(DU(t, Y(\omega), \omega), \theta(t, \omega))$ has a non-vanishing Lyapunov spectrum $\{\cdots < \lambda_{i+1} < \lambda_i < \cdots < \lambda_2 < \lambda_1\}$, viz. $\lambda_i \neq 0$ for all $i \geq 1$.

By the Oseledec theorem (Theorem 2.1.1), the integrability condition in Definition 2.1.2 (a) implies the existence of a discrete Lyapunov spectrum for the linearized cocycle $(DU(t, Y(\omega), \omega), \theta(t, \omega))$ in Definition 2.1.2 (b) above.

The following result is a random version of the saddle point property for hyperbolic linear cocycles. A proof is given in ([Mo.3], Theorem 4, Corollary 2; [M-S.4], Theorem 5.3) within the context of stochastic differential systems with memory; but the arguments therein extend immediately to linear cocycles in Hilbert space.

THEOREM 2.1.2. *(Stable and unstable subspaces)*
Let (T, θ) be a linear cocycle on a Hilbert space H. Assume that $T(t, \omega) : H \to H$ is a compact linear operator for each $t > 0$ and a.a. $\omega \in \Omega$. Suppose that

$$E \log^+ \sup_{0 \leq t_1, t_2 \leq 1} \|T(t_2, \theta(t_1, \cdot))\|_{L(H)} < \infty,$$

and let the cocycle (T, θ) have a non-vanishing Lyapunov spectrum $\{\cdots < \lambda_{i+1} < \lambda_i < \cdots < \lambda_2 < \lambda_1\}$. Pick $i_0 > 1$ such that $\lambda_{i_0} < 0 < \lambda_{i_0 - 1}$.

Then there is a sure event $\Omega^* \in \mathcal{F}$ and stable and unstable subspaces $\{S(\omega), U(\omega) : \omega \in \Omega^*\}$, $\mathcal{F}$-measurable (into the Grassmanian), such that for each $\omega \in \Omega^*$, the following is true:

(i) *$\theta(t, \cdot)(\Omega^*) = \Omega^*$ for all $t \in \mathbf{R}$.*

(ii) *$H = \mathcal{U}(\omega) \oplus \mathcal{S}(\omega)$. The subspace $\mathcal{U}(\omega)$ is finite-dimensional with a fixed non-random dimension, and $\mathcal{S}(\omega)$ is closed with a finite non-random codimension.*

(iii) *(Invariance)*

$$T(t, \omega)(\mathcal{U}(\omega)) = \mathcal{U}(\theta(t, \omega)), \; T(t, \omega)(\mathcal{S}(\omega)) \subseteq \mathcal{S}(\theta(t, \omega)),$$

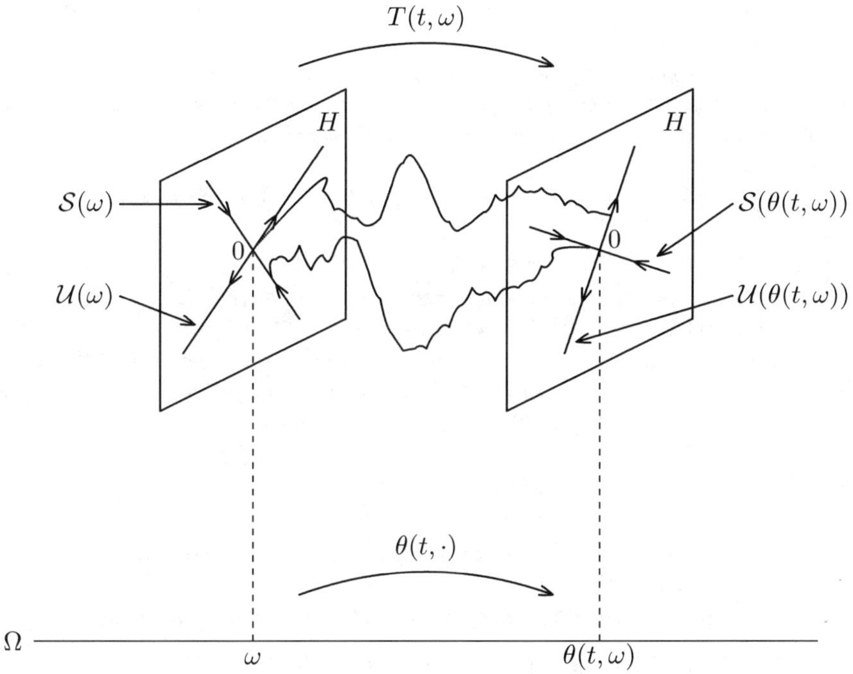

FIGURE 3. The stable and unstable subspaces.

for all $t \geq 0$,

(iv) *(Exponential dichotomies)*

$$|T(t,\omega)(x)| \geq |x|e^{\delta_1 t} \quad \textit{for all} \quad t \geq \tau_1^*, x \in \mathcal{U}(\omega),$$

$$|T(t,\omega)(x)| \leq |x|e^{-\delta_2 t} \quad \textit{for all} \quad t \geq \tau_2^*, x \in \mathcal{S}(\omega),$$

where $\tau_i^* = \tau_i^*(x,\omega) > 0, i = 1,2,$ *are random times and* $\delta_i > 0, i = 1,2,$ *are fixed.*

This theorem is illustrated in Fig. 3.

2.2. The non-linear ergodic theorem

The main objective of this section is to refine and extend discrete-time results of D. Ruelle to the continuous-time setting in Theorem 2.2.1 below. This setting underlies the dynamics of the semilinear see's and spde's studied in Part 1. As will be apparent later, the extension of Ruelle's results to continuous-time is non-trivial. Indeed, Section 2.3 in its entirety is devoted to the proof of Theorem 2.2.1. The main difficulties in the analysis are outlined at the end of this section, after the statement of Theorem 2.2.1.

In the following, denote by $B(x,\rho)$ the open ball in H, radius ρ and center $x \in H$, and by $\bar{B}(x,\rho)$ the corresponding closed ball.

THEOREM 2.2.1. *(The local stable manifold theorem)*
Let (U,θ) be a $C^{k,\epsilon}$ ($k \geq 1, \epsilon \in (0,1]$) perfect cocycle on a separable Hilbert space H such that for each $(t,\omega) \in (0,\infty) \times \Omega$, $U(t,\cdot,\omega): H \to H$ takes bounded sets into relatively compact sets. For any $\rho \in (0,\infty)$, denote by $\|\cdot\|_{k,\epsilon}$ the $C^{k,\epsilon}$-norm on the space $C^{k,\epsilon}(\bar{B}(0,\rho),H)$. Let Y be a hyperbolic stationary point of the cocycle (U,θ) satisfying the following integrability property:

$$\int_\Omega \log^+ \sup_{0\leq t_1,t_2\leq a} \|U(t_2, Y(\theta(t_1,\omega)), \theta(t_1,\omega))\|_{k,\epsilon}\, dP(\omega) < \infty$$

for any fixed $0<\rho, a<\infty$ and $\epsilon \in (0,1]$. Denote by $\{\cdots < \lambda_{i+1} < \lambda_i < \cdots < \lambda_2 < \lambda_1\}$ the Lyapunov spectrum of the linearized cocycle $(DU(t,Y(\omega),\omega), \theta(t,\omega), t\geq 0)$. Define $\lambda_{i_0} := max\{\lambda_i : \lambda_i < 0\}$ if at least one $\lambda_i < 0$. If all finite λ_i are positive, set $\lambda_{i_0} := -\infty$. (Thus λ_{i_0-1} is the smallest positive Lyapunov exponent of the linearized cocycle, if at least one $\lambda_i > 0$; in case all the λ_i's are negative, set $\lambda_{i_0-1} := \infty$.)

Fix $\epsilon_1 \in (0, -\lambda_{i_0})$ and $\epsilon_2 \in (0, \lambda_{i_0-1})$. Then there exist

(i) *a sure event $\Omega^* \in \mathcal{F}$ with $\theta(t,\cdot)(\Omega^*) = \Omega^*$ for all $t \in \mathbf{R}$,*

(ii) *$\bar{\mathcal{F}}$-measurable random variables $\rho_i, \beta_i : \Omega^* \to (0,1)$, $\beta_i > \rho_i > 0$, $i=1,2$, such that for each $\omega \in \Omega^*$, the following is true:*

There are $C^{k,\epsilon}$ ($\epsilon \in (0,1]$) submanifolds $\tilde{\mathcal{S}}(\omega)$, $\tilde{\mathcal{U}}(\omega)$ of $\bar{B}(Y(\omega),\rho_1(\omega))$ and $\bar{B}(Y(\omega),\rho_2(\omega))$ (resp.) with the following properties:

(a) *For $\lambda_{i_0} > -\infty$, $\tilde{\mathcal{S}}(\omega)$ is the set of all $x \in \bar{B}(Y(\omega),\rho_1(\omega))$ such that*

$$|U(n,x,\omega) - Y(\theta(n,\omega))| \leq \beta_1(\omega)\, e^{(\lambda_{i_0}+\epsilon_1)n}$$

for all integers $n \geq 0$. If $\lambda_{i_0} = -\infty$, then $\tilde{\mathcal{S}}(\omega)$ is the set of all $x \in \bar{B}(Y(\omega),\ \rho_1(\omega))$ such that

$$|U(n,x,\omega) - Y(\theta(n,\omega))| \leq \beta_1(\omega)\, e^{\lambda n}$$

for all integers $n \geq 0$ and any $\lambda \in (-\infty,0)$. Furthermore,

$$\limsup_{t\to\infty} \frac{1}{t} \log|U(t,x,\omega) - Y(\theta(t,\omega))| \leq \lambda_{i_0} \tag{2.2.1}$$

for all $x \in \tilde{\mathcal{S}}(\omega)$. Each stable subspace $\mathcal{S}(\omega)$ of the linearized cocycle $(DU(t,Y(\cdot),\cdot),\ \theta(t,\cdot))$ is tangent at $Y(\omega)$ to the submanifold $\tilde{\mathcal{S}}(\omega)$, viz. $T_{Y(\omega)}\tilde{\mathcal{S}}(\omega) = \mathcal{S}(\omega)$. In particular, codim $\tilde{\mathcal{S}}(\omega)$ = codim $\mathcal{S}(\omega)$, is fixed and finite.

(b) $\limsup_{t\to\infty} \frac{1}{t} \log\left[\sup\left\{\frac{|U(t,x_1,\omega) - U(t,x_2,\omega)|}{|x_1 - x_2|} : x_1 \neq x_2,\ x_1, x_2 \in \tilde{\mathcal{S}}(\omega)\right\}\right] \leq \lambda_{i_0}.$

(c) *(Cocycle-invariance of the stable manifolds): There exists $\tau_1(\omega) \geq 0$ such that*

$$U(t,\cdot,\omega)(\tilde{\mathcal{S}}(\omega)) \subseteq \tilde{\mathcal{S}}(\theta(t,\omega)) \tag{2.2.2}$$

for all $t \geq \tau_1(\omega)$. Also

$$DU(t,Y(\omega),\omega)(\mathcal{S}(\omega)) \subseteq \mathcal{S}(\theta(t,\omega)), \quad t \geq 0. \tag{2.2.3}$$

(d) *For $\lambda_{i_0-1} < \infty$, $\tilde{\mathcal{U}}(\omega)$ is the set of all $x \in \bar{B}(Y(\omega),\rho_2(\omega))$ with the property that there is a discrete-time "history" process $y(\cdot,\omega) : \{-n : n \geq 0\} \to H$ such* *that* $y(0,\omega)=x$ *and for each integer* $n\geq 1$, *one has* $U(1,y(-n,\omega),\theta(-n,\omega))=y(-(n-1),\omega)$ *and*

$$|y(-n,\omega)-Y(\theta(-n,\omega))|\leq \beta_2(\omega)e^{-(\lambda_{i_0-1}-\epsilon_2)n}.$$

If $\lambda_{i_0-1}=\infty$, $\tilde{\mathcal{U}}(\omega)$ *is the set of all* $x\in\bar{B}(Y(\omega),\rho_2(\omega))$ *with the property that there is a discrete-time "history" process* $y(\cdot,\omega):\{-n:n\geq 0\}\to H$ *such that* $y(0,\omega)=x$ *and for each integer* $n\geq 1$,

$$|y(-n,\omega)-Y(\theta(-n,\omega))|\leq \beta_2(\omega)e^{-\lambda n},$$

for any $\lambda\in(0,\infty)$. *Furthermore, for each* $x\in\tilde{\mathcal{U}}(\omega)$, *there is a unique continuous-time "history" process also denoted by* $y(\cdot,\omega):(-\infty,0]\to H$ *such that* $y(0,\omega)=x$, $U(t,y(s,\omega),\theta(s,\omega))=y(t+s,\omega)$ *for all* $s\leq 0, 0\leq t\leq -s$, *and*

$$\limsup_{t\to\infty}\frac{1}{t}\log|y(-t,\omega)-Y(\theta(-t,\omega))|\leq -\lambda_{i_0-1}.$$

Each unstable subspace $\mathcal{U}(\omega)$ *of the linearized cocycle* $(DU(t,Y(\cdot),\cdot),\theta(t,\cdot))$ *is tangent at* $Y(\omega)$ *to* $\tilde{\mathcal{U}}(\omega)$, *viz.* $T_{Y(\omega)}\tilde{\mathcal{U}}(\omega)=\mathcal{U}(\omega)$. *In particular,* $\dim\tilde{\mathcal{U}}(\omega)$ *is finite and non-random.*

(e) *Let* $y(\cdot,x_i,\omega), i=1,2$, *be the history processes associated with* $x_i=y(0,x_i,\omega)\in\tilde{\mathcal{U}}(\omega)$, $i=1,2$. *Then*

$$\limsup_{t\to\infty}\frac{1}{t}\log\left[\sup\left\{\frac{|y(-t,x_1,\omega)-y(-t,x_2,\omega)|}{|x_1-x_2|}:x_1\neq x_2,\, x_i\in\tilde{\mathcal{U}}(\omega), i=1,2\right\}\right]$$
$$\leq -\lambda_{i_0-1}.$$

(f) *(Cocycle-invariance of the unstable manifolds):*
There exists $\tau_2(\omega)\geq 0$ *such that*

$$\tilde{\mathcal{U}}(\omega)\subseteq U(t,\cdot,\theta(-t,\omega))(\tilde{\mathcal{U}}(\theta(-t,\omega))) \tag{2.2.4}$$

for all $t\geq\tau_2(\omega)$. *Also*

$$DU(t,\cdot,\theta(-t,\omega))(\mathcal{U}(\theta(-t,\omega)))=\mathcal{U}(\omega),\quad t\geq 0;$$

and the restriction

$$DU(t,\cdot,\theta(-t,\omega))|\mathcal{U}(\theta(-t,\omega)):\mathcal{U}(\theta(-t,\omega))\to\mathcal{U}(\omega),\quad t\geq 0,$$

is a linear homeomorphism onto.

(g) *The submanifolds* $\tilde{\mathcal{U}}(\omega)$ *and* $\tilde{\mathcal{S}}(\omega)$ *are transversal, viz.*

$$H=T_{Y(\omega)}\tilde{\mathcal{U}}(\omega)\oplus T_{Y(\omega)}\tilde{\mathcal{S}}(\omega).$$

Assume, in addition, that the cocycle (U,θ) *is* C^∞. *Then the local stable and unstable manifolds* $\tilde{\mathcal{S}}(\omega)$, $\tilde{\mathcal{U}}(\omega)$ *are also* C^∞.

Fig. 4 summarizes the essential features of the stable manifold theorem.

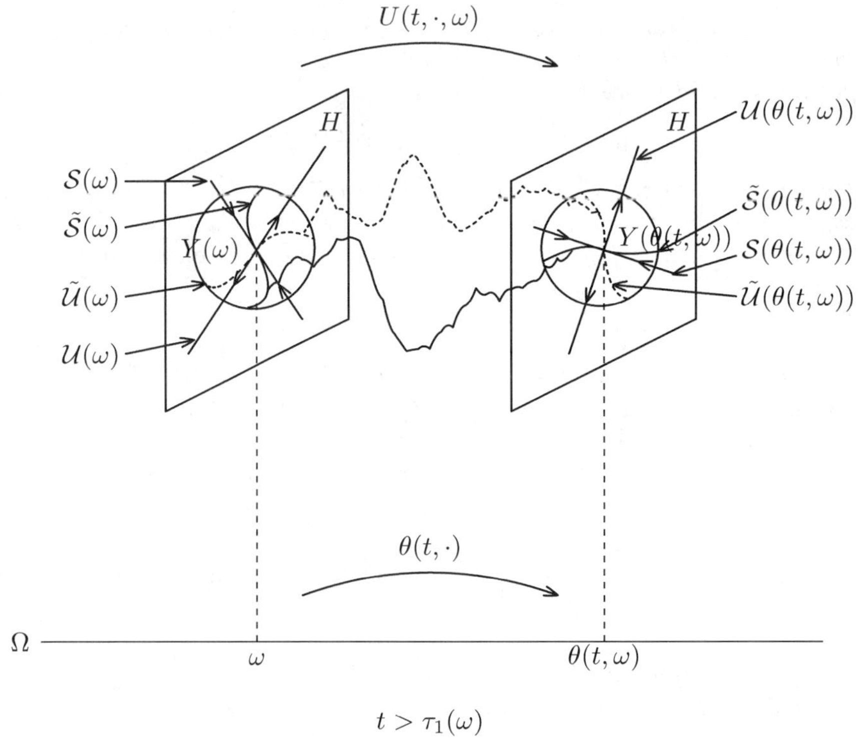

FIGURE 4. The Stable Manifold Theorem.

Before we give a detailed proof of Theorem 2.2.1, we will outline below its basic ingredients.

An outline of the proof of Theorem 2.2.1:

- Since Y is a hyperbolic stationary point of the cocycle (U, θ) (Definition 2.1.3), then the linearized cocycle satisfies the hypotheses of "perfect versions" of the ergodic theorem and Kingman's subadditive ergodic theorem (Lemma 2.3.1 (ii), (iii) in Section 2.3). These refined versions of the ergodic theorems give invariance of the Oseledec spaces under the *continuous-time* linearized cocycle (Theorems 2.1.1-2.1.2). Thus the stable/unstable subspaces will serve as tangent spaces to the local stable/unstable manifolds of the non-linear cocycle (U, θ).
- Define the auxiliary perfect cocycle (Z, θ) by

$$Z(t, \cdot, \omega) := U(t, (\cdot) + Y(\omega), \omega) - Y(\theta(t, \omega)), \ t \in \mathbf{R}^+, \omega \in \Omega.$$

This gives a "centering" of the cocycle around the stationary trajectory $Y(\theta(t))$, with the property that Z has a fixed point at $0 \in H$. Employing the continuous-time integrability estimate in Theorem 2.2.1, the perfect ergodic theorem and the perfect subadditive ergodic theorem, the analysis in ([Ru.2], Theorems 5.1

and 6.1) may be extended to obtain local stable/unstable manifolds for the discrete cocycle $(Z(n,\cdot,\omega),\theta(n,\omega))$ near 0. These manifolds are random objects defined for all ω which are sampled from a $\theta(t,\cdot)$-*invariant sure event* in Ω. The translates of these manifolds by the stationary point $Y(\omega)$ correspond to local stable/unstable manifolds for $U(n,\cdot,\omega)$ near $Y(\omega)$. We then interpolate between discrete times and extend the arguments in [Ru.2] further in order to conclude that the above manifolds for the discrete-time cocycle $(U(n,\cdot,\omega),\theta(n,\omega)), n\geq 1$, also serve as local stable/unstable manifolds for the *continuous-time* cocycle (U,θ) near Y.

- It turns out that the local stable/unstable manifolds are asymptotically invariant under the continuous-time cocycle (U,θ). For the stable manifolds, the invariance follows by arguments based on (a) a refined version of the perfect subadditive ergodic theorem (Lemma 2.3.2, Section 2.3), and (b) difficult estimates using the integrability property of Theorem 2.2.1 and arguments behind the proofs of Ruelle's Theorems 4.1, 5.1 ([Ru.2]). To establish asymptotic invariance of the local unstable manifolds, we introduce the concept of a *stochastic history process* for U, which compensates for the lack of invertibility of the cocycle. Perfection arguments similar to the above give the invariance. This completes the outline of the proof of Theorem 2.2.1.

A full proof of Theorem 2.2.1 will be given in the next section.

2.3. Proof of the local stable manifold theorem

The main objective of this section is to give a proof of Theorem 2.2.1. In particular, we show that the local stable/unstable manifolds for the discrete cocycle are parametrized by sure events which are invariant under the continuous-time shift $\theta(t,\cdot):\Omega\to\Omega$. This is achieved via a number of computations based on perfection techniques. Excursions of the cocycle between discrete times are controlled by integrability hypothesis on the cocycle (U,θ) (Theorem 2.2.1).

"Perfect versions" of the ergodic theorem and Kingman's subadditive ergodic theorem will be used to construct the shift-invariant sure events appearing in the statement of the local stable manifold theorem (Theorem 2.2.1). These results are given in Lemmas 2.3.1 and 2.3.2 below.

The following convention will be frequently used throughout the paper:

DEFINITION 2.3.1. A family of propositions $\{P(\omega):\omega\in\Omega\}$ is said to *hold perfectly in* ω if there is a sure event $\Omega^*\in\mathcal{F}$ such that $\theta(t,\cdot)(\Omega^*)=\Omega^*$ for all $t\in\mathbf{R}$ and $P(\omega)$ is true *for every* $\omega\in\Omega^*$.

LEMMA 2.3.1.

(i) *Let* $\Omega_0\in\bar{\mathcal{F}}$ *be a sure event such that* $\theta(t,\cdot)(\Omega_0)\subseteq\Omega_0$ *for all* $t\geq 0$. *Then there is a sure event* $\Omega_0^*\in\mathcal{F}$ *such that* $\Omega_0^*\subseteq\Omega_0$ *and* $\theta(t,\cdot)(\Omega_0^*)=\Omega_0^*$ *for all* $t\in\mathbf{R}$.

(ii) *Let* $h:\Omega\to\mathbf{R}^+$ *be any function such that there exists an* $\bar{\mathcal{F}}$*-measurable function* $g_1\in L^1(\Omega,\mathbf{R}^+;P)$ *and a sure event* $\Omega_1\in\bar{\mathcal{F}}$ *such that* $\sup_{0\leq u\leq 1} h(\theta(u,\omega))\leq g_1(\omega)$ *for all* $\omega\in\Omega_1$. *Then*

$$\lim_{t\to\infty}\frac{1}{t}h(\theta(t,\omega))=0$$

perfectly in ω.

(iii) *Suppose* $f:\mathbf{R}^+\times\Omega\to\mathbf{R}\cup\{-\infty\}$ *is a process such that for each* $t\in\mathbf{R}^+$, $f(t,\cdot)$ *is* $(\bar{\mathcal{F}},\mathcal{B}(\mathbf{R}\cup\{-\infty\}))$*-measurable and the following conditions hold:*

(a) *There is an $\bar{\mathcal{F}}$-measurable function $g_2 \in L^1(\Omega, \mathbf{R}^+; P)$ and a sure event $\tilde{\Omega}_1 \in \bar{\mathcal{F}}$ such that* $\left[\sup_{0\le u\le 1} f^+(u,\omega) + \sup_{0\le u\le 1} f^+(1-u, \theta(u,\omega))\right] \le g_2(\omega)$ *for all $\omega \in \tilde{\Omega}_1$.*

(b) $f(t_1+t_2,\omega) \le f(t_1,\omega) + f(t_2, \theta(t_1,\omega))$ *for all $t_1, t_2 \ge 0$ and* **all** $\omega \in \Omega$.

Then there is a fixed (non-random) number $f^* \in \mathbf{R} \cup \{-\infty\}$ such that

$$\lim_{t\to\infty} \frac{1}{t} f(t,\omega) = f^*$$

perfectly in ω

PROOF. Assertion (i) is established in Proposition 2.3 ([M-S.2]).

To prove assertions (ii) and (iii) of the lemma, the reader may adapt the proofs of Lemmas 5 and 7 in [Mo.3] and employ assertion (i) above. Cf. also Lemma 3.3 in [M-S.2]. □

Lemma 2.3.2 below is used to construct the *continuous-time* shift-invariant sure events which appear in the statement of Theorem 2.2.1. In essence, the lemma is a continuous-time "perfect version" of Ruelle's Corollary A.2 ([Ru.2], p. 288).

LEMMA 2.3.2. *Assume that the process $f : \mathbf{R}^+ \times \Omega \to \mathbf{R} \cup \{-\infty\}$ is $(\mathcal{B}(\mathbf{R}^+) \otimes \mathcal{F}, \mathcal{B}(\mathbf{R} \cup \{-\infty\}))$-measurable and satisfies the following integrability and subadditivity conditions:*

(a) $\displaystyle\int_\Omega \left[\sup_{0\le t_1, t_2 \le a} f^+(t_1, \theta(t_2,\omega))\right] dP(\omega) < \infty$ *for all* $a \in (0,\infty)$.

(b) $f(t_1+t_2,\omega) \le f(t_1,\omega) + f(t_2, \theta(t_1,\omega))$ *for all $t_1, t_2 \ge 0$ and* **all** $\omega \in \Omega$.

Then there exists a fixed (non-random) $f^* \in \mathbf{R} \cup \{-\infty\}$ such that the following assertions hold perfectly in ω:

(i) $\displaystyle\lim_{t\to\infty} \frac{1}{t} f(t,\omega) = f^*$.

(ii) *Assume $g^* \in \mathbf{R}$ is finite and such that $f^* \le g^*$. Then for each $\epsilon > 0$, there is an $\bar{\mathcal{F}}$-measurable function $K_\epsilon : \Omega \to [0,\infty)$ with the following properties*

$$f(t-s, \theta(s,\omega)) \le (t-s)g^* + \epsilon t + K_\epsilon(\omega), \quad 0 \le s \le t < \infty,$$
$$K_\epsilon(\theta(l,\omega)) \le K_\epsilon(\omega) + \epsilon l, \quad l \in [0,\infty).$$

PROOF. Applying Lemma 2.3.1 (iii), it is easy to see that there is an $f^* \in \mathbf{R} \cup \{-\infty\}$ such that assertion (i) holds for all ω in a sure event $\Omega_2 \in \mathcal{F}$ with $\theta(t,\cdot)(\Omega_2) = \Omega_2$ for all $t \in \mathbf{R}$. The integrability hypotheses (a) and Lemma 2.3.1 (i) imply that there is a sure event $\Omega_0 \subseteq \Omega_2$ such that $\Omega_0 \in \mathcal{F}$, $\theta(t,\cdot)(\Omega_0) = \Omega_0$ for all $t \in \mathbf{R}$, and $\sup_{0\le t_1,t_2\le a} f^+(t_1, \theta(t_2,\omega)) < \infty$ for all $a \ge 0$ and all $\omega \in \Omega_0$. Let g^* be a finite number in $[f^*, \infty)$. Define the non-negative process $g : \mathbf{R}^+ \times \Omega \to \mathbf{R}^+$ by

$$g(t,\omega) := \begin{cases} \max\{f(t,\omega) - tg^*, 0\}, & t \ge 0,\ \omega \in \Omega_0, \\ 0 & t \ge 0,\ \omega \notin \Omega_0. \end{cases}$$

Then g is $(\mathcal{B}(\mathbf{R}^+) \otimes \mathcal{F}, \mathcal{B}(\mathbf{R}^+))$-measurable and satisfies conditions (a) and (b).

Now consider the non-negative process $g' : \mathbf{R}^+ \times \Omega \to \mathbf{R}^+$ defined by

$$g'(t,\omega) := \sup_{0\le s\le t} [g(s,\omega) + g(t-s, \theta(s,\omega))], \quad t \ge 0,\ \omega \in \Omega.$$

Observe that the projection of a $(\mathcal{B}(\mathbf{R}^+)\otimes\mathcal{F})$-measurable set is $\bar{\mathcal{F}}$-measurable ([Co], p. 281). Therefore, g' satisfies all the hypotheses of Lemma 2.3.1 (iii). This gives a non-negative g'^* such that $\lim_{t\to\infty}\frac{1}{t}g'(t,\omega) = g'^*$ for all ω in a sure event $\Omega_3 \in \mathcal{F}$, with $\theta(t,\cdot)(\Omega_3) = \Omega_3$ for all $t \in \mathbf{R}$.

We will show next the following convergence in probability:

$$\lim_{t\to\infty}\frac{1}{t}\sup_{0\le s\le t} g(t-s,\theta(s,\cdot)) = 0. \tag{2.3.1}$$

To do this, observe that the process $h : \mathbf{R}^+ \times \Omega \to \mathbf{R}$, $h(t,\omega) := g(t,\theta(-t,\omega))$, $t \in \mathbf{R}^+$, $\omega \in \Omega$, satisfies the conditions of Lemma 2.3.1 (iii). Therefore,

$$\lim_{t\to\infty}\frac{1}{t}h(t,\cdot) = 0$$

almost surely and hence in probability. Pick $\delta, t_0 > 0$ such that $P(\frac{1}{t}h(t,\cdot) \ge \delta) < \delta$ for all $t \ge t_0$. Let $t \ge t_0$. Then

$$\begin{aligned}\sup_{0\le s\le t}\frac{1}{t}g(t-s,\theta(s,\omega)) &\le \sup_{0\le s\le t-t_0}\frac{1}{t}g(t-s,\theta(s,\omega)) + \sup_{t-t_0\le s\le t}\frac{1}{t}g(t-s,\theta(s,\omega))\\ &\le \sup_{0\le s\le t-t_0}\frac{1}{t}g(t-s,\theta(-(t-s),\theta(t,\omega)))\\ &\qquad\qquad + \sup_{t-t_0\le s\le t}\frac{1}{t}g(t-s,\theta(s,\omega)).\end{aligned}$$

By condition (a), the second term in the right hand side of the last inequality converges to zero in probability. The probability that the first term is less than or equal to δ is at least $1-\delta$. Hence (2.3.1) holds.

It follows easily from (2.3.1) that $g'^* = 0$. This implies that assertion (i) holds for all ω in a sure event $\Omega_4 \in \mathcal{F}$ with $\Omega_4 \subseteq \Omega_0 \cap \Omega_3$ and $\theta(t,\cdot)(\Omega_4) = \Omega_4$ for all $t \in \mathbf{R}$. To complete the proof of assertion (ii), let $\epsilon > 0$ and define the $(\bar{\mathcal{F}}, \mathcal{B}(\mathbf{R}^+))$-measurable function $K_\epsilon : \Omega_4 \to [0,\infty)$ by

$$K_\epsilon(\omega) := \sup_{0\le s\le t<\infty}[g(t-s,\theta(s,\omega)) - \epsilon t]$$

for all $\omega \in \Omega_4$. This completes the proof of the lemma. □

Lemma 2.3.3 below is essentially a "perfect version" of Proposition 3.2 in [Ru.2], p. 257. Our Lemma 2.3.2 plays a crucial role in the proof of Lemma 2.3.3.

In the statement of the following lemma, we will use $\mathcal{B}_s(L(H))$ to denote the Borel σ-algebra on $L(H)$ generated by the strong topology on $L(H)$, viz. the smallest topology on $L(H)$ for which all evaluations $L(H) \ni A \mapsto A(z) \in H, z \in H$, are continuous.

LEMMA 2.3.3. *Suppose* $(T^t(\omega), \theta(t,\omega))$, $t \ge 0$, *is a perfect cocycle of bounded linear operators in* H *satisfying the following hypotheses:*

(i) *The process* $\mathbf{R}^+ \times \Omega \ni (t,\omega) \mapsto T^t(\omega) \in L(H)$ *is* $(\mathcal{B}(\mathbf{R}^+)\otimes\mathcal{F}, \mathcal{B}_s(L(H)))$-*measurable.*

(ii) *The map* $\mathbf{R}^+ \times \Omega \ni (t,\omega) \mapsto \theta(t,\omega) \in \Omega$ *is* $(\mathcal{B}(\mathbf{R}^+)\otimes\mathcal{F}, \mathcal{F})$-*measurable, and is a group of ergodic* P-*preserving transformations on* $(\Omega, \mathcal{F}, P)$.

(iii) $E\sup_{0\le t_1,t_2\le a}\log^+\|T^{t_2}(\theta(t_1,\cdot))\|_{L(H)} < \infty$ *for any finite* $a > 0$.

(iv) For each $t > 0$, $T^t(\omega)$ is compact, perfectly in ω.

(v) *For any $u \in H$, the map $[0,\infty) \ni t \mapsto T^t(\omega)(u) \in H$ is continuous, perfectly in ω.*

Let $\{\cdots < \lambda_{i+1} < \lambda_i < \cdots < \lambda_2 < \lambda_1\}$ be the Lyapunov spectrum of $(T^t(\omega), \theta(t,\omega))$, with Oseledec spaces

$$\cdots E_{i+1}(\omega) \subset E_i(\omega) \subset \cdots \subset E_2(\omega) \subset E_1(\omega) = H.$$

Let $j_0 \geq 1$ be any fixed integer with $\lambda_{j_0} > -\infty$. Let the integer function $r : \{1, 2, \cdots, Q\} \to \{1, 2, \cdots, j_0\}$ "count" the multiplicities of the Lyapunov exponents in the sense that $r(1) = 1$, $r(Q) = j_0$, and for each $1 \leq i \leq j_0$, the number of integers in $r^{-1}(i)$ is the multiplicity of λ_i. Set $V_n(\omega) := E_{j_0+1}(\theta(n,\omega)), n \geq 0$.

Then the sequence $T_n(\omega) := T^1(\theta((n-1),\omega)), n \geq 1$, satisfies Condition (S) *of* ([Ru.2], pp. 256-257) *perfectly in ω with $Q = \operatorname{codim} E_{j_0+1}(\omega)$. In particular, there is an $\mathcal{F}$-measurable set of Q orthonormal vectors $\{\xi_0^{(1)}(\omega), \cdots, \xi_0^{(Q)}(\omega)\}$ such that $\xi_0^{(k)}(\omega) \in [E_{r(k)}(\omega) \backslash E_{r(k)+1}(\omega)]$ for $k = 1, \cdots, Q$, perfectly in ω, and satisfying the following properties:*

Set $\xi_t^{(k)}(\omega) := \dfrac{T^t(\omega)(\xi_0^{(k)}(\omega))}{|T^t(\omega)(\xi_0^{(k)}(\omega))|}$, *and for any $u \in H$, write*

$$u = \sum_{k=1}^{Q} u_t^{(k)}(\omega)\xi_t^{(k)}(\omega) + u_t^{(Q+1)}(\omega), \quad u_t^{(Q+1)}(\omega) \in V_0(\theta(t,\omega)), \ \omega \in \Omega.$$

Then for any $\epsilon > 0$, there is an $\bar{\mathcal{F}}$-measurable random constant $D_\epsilon(\omega) > 0$ such that the following inequalities hold perfectly in ω:

$$\begin{aligned} |u_t^{(k)}(\omega)| &\leq D_\epsilon(\omega)e^{\epsilon t}|u| \\ |u_t^{(Q+1)}(\omega)| &\leq D_\epsilon(\omega)e^{\epsilon t}|u| \\ D_\epsilon(\theta(l,\omega)) &\leq D_\epsilon(\omega)e^{\epsilon l} \end{aligned}$$

for all $t \geq 0$, $1 \leq k \leq Q$ and for all $l \in [0,\infty)$.

Furthermore, all the random constants in Ruelle's Condition (S) ([Ru.2], pp. 256-257) *may be chosen to be $\bar{\mathcal{F}}$-measurable in ω.*

PROOF. Our proof runs along similar lines to that of Proposition 3.2 in [Ru.2]: However, one has to maintain the non-trivial requirement that all relevant arguments hold perfectly in ω.

It is assumed throughout this proof that the reader is familiar with Ruelle's conditions (S): (S1)-(S4) as spelled out in ([Ru.2], pp. 256-257).

Observe first that $T_n(\omega)$ satisfies (S1) perfectly in ω. This holds because of (iii), the perfect cocycle property, Lemma 2.3.1 and the proof of Theorem 4 ([Mo.1]). Note that, by the ordering of the fixed Lyapunov spectrum, relation (3.4) of [Ru.2] holds perfectly. Denote by Ω^* the $\theta(t,\cdot)$-invariant sure event where (S1) holds. Using ergodicity of θ and the fact that $\operatorname{codim} V_0(\omega) = Q$, for all $\omega \in \Omega^*$, it follows that $\operatorname{codim} V_n(\omega) = \operatorname{codim} E_{j_0+1}(\theta(n,\omega)) = Q$. Therefore, (S2) is satisfied for all $\omega \in \Omega^*$.

We next prove that (S3) holds perfectly. To do this, we will prove the stronger assertion that the *continuous-time* cocycle $(T^t(\omega), \theta(t,\omega))$ satisfies (S3) perfectly in

ω. Set $\hat{T}^t(\omega) := T^t(\omega)|V_0(\omega), \omega \in \Omega^*, t \geq 0$. Hence $\hat{T}^t(\omega)(V_0(\omega)) \subseteq V_0(\theta(t,\omega))$, and the following cocycle identity

$$\hat{T}^{t_1+t_2}(\omega) = \hat{T}^{t_2}(\theta(t_1,\omega)) \circ \hat{T}^{t_1}(\omega)$$

holds for all $\omega \in \Omega^*, t \geq 0$. Denote $F_t(\omega) := \log \|\hat{T}^t(\omega)\|, \ \omega \in \Omega^*, t \geq 0$. Hypothesis (iii) of the lemma easily implies that $E \sup_{0 \leq t_1, t_2 \leq a} F_{t_2}^+(\theta(t_1, \cdot)) < \infty$ for any finite $a > 0$. Furthermore, $(F_t(\omega), \theta(t,\omega))$ is perfectly subadditive because of the above cocycle identity. Applying Lemma 2.3.1, we obtain a fixed number $F^* \in \mathbf{R} \cup \{-\infty\}$ such that

$$\lim_{t \to \infty} \frac{1}{t} F_t(\omega) = F^*$$

perfectly in ω. Suppose $S = j_0$. When $\lambda_{j_0+1} > -\infty$, set $\mu^{(S+1)} := \lambda_{j_0+1}$; and when $\lambda_{j_0+1} = -\infty$, take $\mu^{(S+1)}$ to be any fixed number in $(-\infty, \lambda_{j_0})$. Using (3.5), p. 257 of [Ru.2], it follows that $F^* \leq \mu^{(S+1)}$. Suppose $\epsilon > 0$ and $\lambda_{j_0+1} > -\infty$. Then by Lemma 2.3.2(ii), there is an $\bar{\mathcal{F}}$-measurable function $K_\epsilon : \Omega \to [0, \infty)$ such that

$$\log \|\hat{T}^{t-s}(\theta(s,\omega))\| \leq (t-s)\mu^{(S+1)} + \epsilon t + K_\epsilon(\omega), \quad 0 \leq s \leq t < \infty, \tag{2.3.2}$$

and

$$K_\epsilon(\theta(l,\omega)) \leq K_\epsilon(\omega) + \epsilon l, \quad l \in [0, \infty),$$

perfectly in ω. When $\lambda_{j_0+1} = -\infty$, the inequality (2.3.2) holds where $\mu^{(S+1)}$ is replaced by any (finite) number in $(-\infty, \lambda_{j_0})$. Now let m, n be positive integers such that $m < n$. In (2.3.2), replace t by n and s by $m+1$ to see that $T_n(\omega)$, $n \geq 1$, satisfies (S3) perfectly in ω.

The rest of this proof will now focus on showing that the sequence $T_n(\omega)$, $n \geq 1$, also satisfies Ruelle's condition (S4) perfectly in ω. Indeed, we will establish the stronger statement that the continuous-time cocycle $(T^t(\omega), \theta(t,\omega))$ satisfies (S4) perfectly in ω. Using the orthogonal decomposition $H = V_0(\theta(t,\omega)) \oplus V_0(\theta(t,\omega))^\perp$, write

$$T^t(\omega)(\xi) = \check{T}^t(\omega)(\xi) + \tilde{T}^t(\omega)(\xi), \quad \xi \in H, t \geq 0, \omega \in \Omega^*. \tag{2.3.3}$$

That is, $\tilde{T}^t(\omega)(\xi) \in V_0(\theta(t,\omega))$ and $\check{T}^t(\omega)(\xi) \in V_0(\theta(t,\omega))^\perp$ are the orthogonal projections of $T^t(\omega)(\xi)$ on $V_0(\theta(t,\omega))$ and $V_0(\theta(t,\omega))^\perp$, respectively. Thus (2.3.3) defines a family of continuous linear operators $\check{T}^t(\omega) : H \to V_0(\theta(t,\omega))^\perp \subseteq H$, $\tilde{T}^t(\omega) : H \to V_0(\theta(t,\omega)) \subseteq H$, $t \geq 0$. We now show that the family $(\check{T}^t(\omega), \theta(t,\omega)), \omega \in \Omega$, satisfies the perfect cocycle property in $L(H)$. To prove this, we fix any $\omega \in \Omega$. Then by the cocycle property of $(T^t(\omega), \theta(t,\omega))$ and (2.3.3), we obtain

$$\begin{aligned} T^{t_1+t_2}(\omega)(\xi) &= T^{t_2}(\theta(t_1,\omega))[T^{t_1}(\omega)(\xi)] \\ &= \check{T}^{t_2}(\theta(t_1,\omega))[\check{T}^{t_1}(\omega)(\xi)] + \check{T}^{t_2}(\theta(t_1,\omega))[\tilde{T}^{t_1}(\omega)(\xi)] \\ &\quad + \tilde{T}^{t_2}(\theta(t_1,\omega))[\check{T}^{t_1}(\omega)(\xi)] + \tilde{T}^{t_2}(\theta(t_1,\omega))[\tilde{T}^{t_1}(\omega)(\xi)]. \end{aligned} \tag{2.3.4}$$

for all $t_1, t_2 \geq 0$, $\xi \in H$. Furthermore, $\check{T}^t(\omega)(\xi) = 0$ for all $\xi \in V_0(\omega)$, because $V_0(\omega)$ is invariant under the cocycle $(T^t(\omega), \theta(t,\omega))$. Thus, $\check{T}^{t_2}(\theta(t_1,\omega))[\tilde{T}^{t_1}(\omega)(\xi)] = 0$ for all $\xi \in H$, and (2.3.4) yields

$$\begin{aligned} &T^{t_1+t_2}(\omega)(\xi) \\ &= \check{T}^{t_2}(\theta(t_1,\omega))[\check{T}^{t_1}(\omega)(\xi)] + \tilde{T}^{t_2}(\theta(t_1,\omega))[\check{T}^{t_1}(\omega)(\xi)] + \\ &\quad + \tilde{T}^{t_2}(\theta(t_1,\omega))[\tilde{T}^{t_1}(\omega)(\xi)], \quad t_1, t_2 \geq 0, \ \xi \in H. \end{aligned} \tag{2.3.5}$$

Now

$$T^{t_1+t_2}(\omega)(\xi) = \check{T}^{t_1+t_2}(\omega)(\xi) + \tilde{T}^{t_1+t_2}(\omega)(\xi) \tag{2.3.6}$$

for all $\xi \in H$. In the right hand side of (2.3.5), the first term belongs to $V_0(\theta(t_1 + t_2, \omega))^\perp$, while the second two terms belong to $V_0(\theta(t_1 + t_2, \omega))$. So by uniqueness of the orthogonal decomposition, it follows from (2.3.6) and (2.3.5) that

$$\check{T}^{t_1+t_2}(\omega)(\xi) = \check{T}^{t_2}(\theta(t_1, \omega))[\check{T}^{t_1}(\omega)(\xi)] \tag{2.3.7}$$

for all $\xi \in H$. Hence $(\check{T}^t(\omega), \theta(t, \omega))$ is a perfect cocycle in $L(H)$.

We next verify that both cocycles $(T^t(\omega), \theta(t, \omega))$ and $(\check{T}^t(\omega), \theta(t, \omega))$ satisfy the conditions of the perfect Oseledec theorem (Theorem 2.1.1). To see this, note that

$$E \sup_{0 \le t_1, t_2 \le a} \log^+ \|\check{T}^{t_2}(\theta(t_1, \cdot))\|_{L(H)} < \infty \tag{2.3.8}$$

for any finite $a > 0$. This follows immediately from the integrability property (iii) of the lemma. Now apply Theorem 2.1.1 to $(T^t(\omega), \theta(t, \omega))$ and $(\check{T}^t(\omega), \theta(t, \omega))$. This gives the following limits

$$\lim_{t \to \infty} \frac{1}{t} \log |\check{T}^t(\omega)(\xi)| = \check{l}_\xi, \qquad \lim_{t \to \infty} \frac{1}{t} \log |T^t(\omega)(\xi)| = l_\xi$$

perfectly in ω for all $\xi \in H$, with $l_\xi, \check{l}_\xi$ fixed numbers in $\mathbf{R} \cup \{-\infty\}$. We now apply (3.6) in ([Ru.2], p. 259) to obtain

$$\check{l}_\xi = \lim_{n \to \infty} \frac{1}{n} \log |\check{T}^n(\omega)(\xi)| = \lim_{n \to \infty} \frac{1}{n} \log |T^n(\omega)(\xi)| = l_\xi$$

for a.a. ω and all $\xi \in H \backslash V_0(\omega)$. Therefore the equality

$$\lim_{t \to \infty} \frac{1}{t} \log |\check{T}^t(\omega)(\xi)| = \lim_{t \to \infty} \frac{1}{t} \log |T^t(\omega)(\xi)|$$

holds perfectly in ω for all $\xi \in H \backslash V_0(\omega)$. Hence, relation (3.6) in ([Ru.2], p. 259) may be replaced by the continuous-time "perfect" relation

$$\lim_{t \to \infty} \frac{1}{t} \log \frac{|\check{T}^t(\omega)(\xi)|}{|T^t(\omega)(\xi)|} = 0 \tag{2.3.9}$$

for all $\xi \in H \backslash V_0(\omega)$.

By ([C-V], Theorem III.6, p. 65) and Gram-Schmidt orthogonalization, we may select a set of Q, $\mathcal{F}$-measurable, orthonormal vectors $\{\xi_0^{(1)}(\omega), \cdots, \xi_0^{(Q)}(\omega)\}$ such that $\xi_0^{(k)}(\omega) \in [E_{r(k)}(\omega) \backslash E_{r(k)+1}(\omega)] \cap V_0(\omega)^\perp$ for $k = 1, \cdots, Q$, perfectly in ω. In the argument in [Ru.2], p. 259, replace (3.6) by (2.3.9) above, n by t, $\xi_n^{(k)}$ by $\xi_t^{(k)}(\omega) := \dfrac{T^t(\omega)(\xi_0^{(k)}(\omega))}{|T^t(\omega)(\xi_0^{(k)}(\omega))|}$, V_n by $V_0(\theta(t, \omega))$, and $\eta_n^{(k)}$ by $\eta_t^{(k)}(\omega) := \dfrac{\check{T}^t(\omega)(\xi_0^{(k)}(\omega))}{|T^t(\omega)(\xi_0^{(k)}(\omega))|}$. Therefore for $u \in H$, we write

$$u = \sum_{k=1}^{Q} u_t^{(k)}(\omega) \xi_t^{(k)}(\omega) + u_t^{(Q+1)}(\omega), \quad u_t^{(Q+1)}(\omega) \in V_0(\theta(t, \omega)), \tag{2.3.10}$$

perfectly in ω for all $t \geq 0$. Furthermore, as in [Ru.2], p. 259, (2.3.9) implies that

(2.3.11)
$$\lim_{t\to\infty} \frac{1}{t} \log|\det(\eta_t^{(1)}(\omega), \cdots, \eta_t^{(Q)}(\omega))| = 0,$$

perfectly in ω.

It remains to prove that for each $\epsilon > 0$, there is an $\bar{\mathcal{F}}$-measurable non-negative function $D_\epsilon : \Omega \to (0, \infty)$ such that the following inequalities

(2.3.12)
$$\left.\begin{aligned} |u_t^{(k)}(\omega)| &\leq D_\epsilon(\omega)e^{\epsilon t}|u| \\ |u_t^{(Q+1)}(\omega)| &\leq D_\epsilon(\omega)e^{\epsilon t}|u| \\ D_\epsilon(\theta(l,\omega)) &\leq D_\epsilon(\omega)e^{\epsilon l} \end{aligned}\right\}$$

hold perfectly in ω, for all $t \geq 0$, $1 \leq k \leq Q$ and for all $l \in [0, \infty)$.

In order to establish the inequalities (2.3.12), we define

(2.3.13)
$$D_\epsilon(\omega) := 1 + Q \cdot \sup_{0\leq s\leq t<\infty} e^{-\epsilon t} |\det(\eta_{t-s}^{(1)}(\theta(s,\omega)), \eta_{t-s}^{(2)}(\theta(s,\omega)), \cdots, \eta_{t-s}^{(Q)}(\theta(s,\omega)))|^{-1}$$

perfectly in ω.

First of all we must show that $D_\epsilon(\omega)$ is finite perfectly in ω. Surprisingly, this will require some work. Let $0 \leq s \leq t$. Observe that the determinant of the linear operator $\check{T}^{t-s}(\theta(s,\omega))$ is given by $\dfrac{|\wedge_{k=1}^{Q} \check{T}^{t-s}(\theta(s,\omega))(v_k)|}{|\wedge_{k=1}^{Q} v_k|}$ for *any choice of basis* $\{v_1, \cdots, v_Q\}$ in $V_0(\theta(s,\omega))^\perp$. Therefore, the following inequalities hold perfectly in ω:

$$\begin{aligned}
&|\det(\eta_{t-s}^{(1)}(\theta(s,\omega)), \cdots, \eta_{t-s}^{(Q)}(\theta(s,\omega)))|^{-1} \\
&= \frac{\Pi_{k=1}^{Q} |T^{t-s}(\theta(s,\omega))(\xi_0^{(k)}(\theta(s,\omega)))|}{|\det(\check{T}^{t-s}(\theta(s,\omega))(\xi_0^{(1)}(\theta(s,\omega))), \cdots, \check{T}^{t-s}(\theta(s,\omega))(\xi_0^{(Q)}(\theta(s,\omega))))|} \\
&= \frac{\Pi_{k=1}^{Q} [|T^{t-s}(\theta(s,\omega))(\xi_0^{(k)}(\theta(s,\omega)))|] \cdot |\wedge_{k=1}^{Q} [\check{T}^s(\omega)(\xi_0^{(k)}(\omega))]|}{|\det(\check{T}^{t-s}(\theta(s,\omega))(\check{T}^s(\omega)(\xi_0^{(1)}(\omega))), \cdots, \check{T}^{t-s}(\theta(s,\omega))(\check{T}^s(\omega)(\xi_0^{(Q)}(\omega))))|} \\
&\leq \frac{\Pi_{k=1}^{Q} [|T^{t-s}(\theta(s,\omega))(\xi_0^{(k)}(\theta(s,\omega)))| \cdot |\check{T}^s(\omega)(\xi_0^{(k)}(\omega))|]}{|\det(\check{T}^t(\omega)(\xi_0^{(1)}(\omega)), \cdots, \check{T}^t(\omega)(\xi_0^{(Q)}(\omega)))|}
\end{aligned}$$

(2.3.14)
$$\begin{aligned}
&= \frac{\Pi_{k=1}^{Q} [|T^{t-s}(\theta(s,\omega))(\xi_0^{(k)}(\theta(s,\omega)))| \cdot |\check{T}^s(\omega)(\xi_0^{(k)}(\omega))|]}{\|[\check{T}^t(\omega)|V_0(\omega)^\perp]^{\wedge Q}\|} \\
&\leq \frac{\|T^{t-s}(\theta(s,\omega))\|^Q \cdot \|\check{T}^s(\omega)\|^Q}{\|[\check{T}^t(\omega)|V_0(\omega)^\perp]^{\wedge Q}\|}.
\end{aligned}$$

By the integrability condition (iii), it follows that

(2.3.15)
$$\sup_{0\leq s\leq t\leq a} \|T^{t-s}(\theta(s,\omega))\|^Q \cdot \|\check{T}^s(\omega)\|^Q < \infty,$$

perfectly in ω for any finite $a > 0$.

We now prove that for each finite $a > 0$,

(2.3.16)
$$\sup_{0\leq s\leq t\leq a} |\det(\eta_{t-s}^{(1)}(\theta(s,\omega)), \cdots, \eta_{t-s}^{(Q)}(\theta(s,\omega)))|^{-1} < \infty$$

perfectly in ω. To see this, define the compact set

$$S(\omega) :=\{(t, v_1, \cdots, v_Q) : t \in [0, a], v_k \in V_0(\omega)^{\perp}, \\ |v_k| = 1, < v_k, v_l >= 0, 1 \leq k < l \leq Q\}$$

for $\omega \in \Omega$. Thus (2.3.16) will hold if we prove that

$$\inf_{(t,v_1,\cdots,v_Q)\in S(\omega)} |\wedge_{k=1}^{Q} [\check{T}^t(\omega)(v_k)]| > 0 \tag{2.3.17}$$

perfectly in ω.

To prove (2.3.17), we observe that each map

$$\check{T}^t(\omega)|V_0(\omega)^{\perp} : V_0(\omega)^{\perp} \to V_0(\theta(t, \omega))^{\perp}$$

is injective for each $t \geq 0$ perfectly in ω. This is an easy consequence of the cocycle property and the fact that $\lambda_{j_0} > -\infty$. In fact,

$$|\wedge_{k=1}^{Q} [\check{T}^t(\omega)(v_k)]| > 0 \tag{2.3.18}$$

for all $(t, v_1, \cdots, v_Q) \in S(\omega)$. Furthermore, the map

$$[0, a] \times [V_0(\omega)^{\perp}]^Q \ni (t, v_1, \cdots, v_Q) \mapsto |\wedge_{k=1}^{Q} [\check{T}^t(\omega)(v_k)]| \in [0, \infty)$$

is jointly continuous, by hypothesis (v) of the lemma. By compactness of $S(\omega)$, (2.3.18) implies (2.3.17). Therefore, (2.3.16) follows from (2.3.14), (2.3.15) and (2.3.17).

The following convergence

$$\lim_{t\to\infty} \frac{1}{t} \log \sup_{0\leq s\leq t} |\det(\eta_{t-s}^{(1)}(\theta(s, \omega)), \cdots, \eta_{t-s}^{(Q)}(\theta(s, \omega)))|^{-1} = 0 \tag{2.3.19}$$

holds perfectly in ω. To prove this convergence, note that (2.3.14) implies the following estimate

$$|\det(\eta_{t-s}^{(1)}(\theta(s, \omega)), \cdots, \eta_{t-s}^{(Q)}(\theta(s, \omega)))|^{-1} \\ \leq \frac{\Pi_{k=1}^{Q}\{\|[T^{t-s}(\theta(s, \omega))|E_{r(k)}(\theta(s, \omega)]\| \cdot \|[\check{T}^s(\omega)|E_{r(k)}(\omega)]\|\}}{\|[\check{T}^t(\omega)|V_0(\omega)^{\perp}]^{\wedge Q}\|}$$

for $0 \leq s \leq t$ perfectly in ω. Let $\epsilon > 0$ be arbitrary. Taking $\frac{1}{t} \log \sup\limits_{0\leq s\leq t}$ on both sides of the above inequality and applying Lemma 2.3.2(ii) yields the following inequalities

$$\frac{1}{t} \log \sup_{0\leq s\leq t} |\det(\eta_{t-s}^{(1)}(\theta(s, \omega)), \cdots, \eta_{t-s}^{(Q)}(\theta(s, \omega)))|^{-1}$$

$$\leq \frac{1}{t} \sup_{0\leq s\leq t} \left\{ \sum_{k=1}^{Q} (\log \|[T^{t-s}(\theta(s, \omega))|E_{r(k)}(\theta(s, \omega)]\| + \log \|[\check{T}^s(\omega)|E_{r(k)}(\omega)]\|) \right\} \\ - \frac{1}{t} \log \|[\check{T}^t(\omega)|V_0(\omega)^{\perp}]^{\wedge Q}\|$$

$$\leq \frac{1}{t} \sup_{0\leq s\leq t} \left\{ \sum_{k=1}^{Q} (t - s)\lambda_{r(k)} + \epsilon t + K_{\epsilon}^1(\omega) + \sum_{k=1}^{Q} s\lambda_{r(k)} + \epsilon s + K_{\epsilon}^2(\omega) \right\}$$

$$
\begin{aligned}
&\quad - \frac{1}{t} \log \|[\check{T}^t(\omega)|V_0(\omega)^\perp]^{\wedge Q}\| \\
&= \sum_{k=1}^{Q} \lambda_{r(k)} + 2\epsilon + \frac{1}{t}[K_\epsilon^1(\omega) + K_\epsilon^2(\omega)] - \frac{1}{t} \log \|[\check{T}^t(\omega)|V_0(\omega)^\perp]^{\wedge Q}\|,
\end{aligned}
$$

for $t > 0$, perfectly in ω, with $K_\epsilon^i(\omega), i = 1, 2$, finite positive random constants independent of t. Therefore, the above inequality implies that

$$
\begin{aligned}
\limsup_{t\to\infty} \frac{1}{t} \log \sup_{0\le s\le t} &|\det(\eta_{t-s}^{(1)}(\theta(s,\omega)), \cdots, \eta_{t-s}^{(Q)}(\theta(s,\omega)))|^{-1} \\
&\le \sum_{k=1}^{Q} \lambda_{r(k)} + 2\epsilon - \liminf_{t\to\infty} \frac{1}{t} \log \|[\check{T}^t(\omega)|V_0(\omega)^\perp]^{\wedge Q}\| \\
&= \sum_{k=1}^{Q} \lambda_{r(k)} + 2\epsilon - \sum_{k=1}^{Q} \lambda_{r(k)} \\
&= 2\epsilon.
\end{aligned}
$$

Since $\epsilon > 0$ is arbitrary, then

$$
\limsup_{t\to\infty} \frac{1}{t} \log \sup_{0\le s\le t} |\det(\eta_{t-s}^{(1)}(\theta(s,\omega)), \cdots, \eta_{t-s}^{(Q)}(\theta(s,\omega)))|^{-1} \le 0 \tag{2.3.20}
$$

perfectly in ω. The convergence (2.3.11) immediately implies the inequality

$$
\begin{aligned}
\liminf_{t\to\infty} \frac{1}{t} \log \sup_{0\le s\le t} &|\det(\eta_{t-s}^{(1)}(\theta(s,\omega)), \cdots, \eta_{t-s}^{(Q)}(\theta(s,\omega)))|^{-1} \\
&\ge \liminf_{t\to\infty} \frac{1}{t} \log |\det(\eta_t^{(1)}(\omega), \cdots, \eta_t^{(Q)}(\omega))|^{-1} = 0
\end{aligned} \tag{2.3.21}
$$

Thus (2.3.19) follows from (2.3.20) and (2.3.21).

From (2.3.16), (2.3.19) and (2.3.13), we conclude that $D_\epsilon(\omega)$ is finite perfectly in ω.

From Definition 2.3.13 of $D(\omega)$, one immediately gets the last inequality in (2.3.12).

It remains to show the first two inequalities in (2.3.12). In the right hand side of (2.3.10), we look at the terms

$$
\check{u}(\omega) = \sum_{k=1}^{Q} u_t^{(k)}(\omega)\eta_t^{(k)}(\omega), \quad u \in H, t \ge 0,
$$

where $\check{u}(\omega)$, $\eta_t^{(k)}(\omega), 1 \le k \le Q$, are viewed as column vectors in $\mathbf{R}^Q$ with respect to the basis $\{\xi_0^{(k)}(\theta(t,\omega)) : 1 \le k \le Q\}$. Using Cramer's rule, the above equation may be solved for each $u_t^{(k)}(\omega)$. In view of (2.3.13), this yields the following estimates

$$\begin{aligned}
|u_t^{(k)}(\omega)| &= \left|\frac{\det(\eta_t^{(1)}(\omega),\cdots,\eta_t^{(k-1)}(\omega),\tilde{u}(\omega),\eta_t^{(k+1)}(\omega),\cdots,\eta_t^{(Q)}(\omega))}{\det(\eta_t^{(1)}(\omega),\cdots,\eta_t^{(Q)}(\omega))}\right| \\
&\le \frac{|\tilde{u}(\omega)|}{|\det(\eta_t^{(1)}(\omega),\cdots,\eta_t^{(Q)}(\omega))|} \\
&\le \frac{[D_\epsilon(\omega)-1]}{Q}|u|e^{\epsilon t} \\
&\le D_\epsilon(\omega)|u|e^{\epsilon t}, \quad 1\le k\le Q,\, t\ge 0,
\end{aligned} \tag{2.3.22}$$

perfectly in ω. By virtue of (2.3.10), the triangle inequality and (2.3.22), one gets

$$|u_t^{(Q+1)}(\omega)| \le |u| + \sum_{k=1}^{Q} |u_t^{(k)}(\omega)| \le D_\epsilon(\omega)|u|e^{\epsilon t}, \quad t\ge 0,$$

perfectly in ω. Therefore, $T_n(\omega)$ satisfies (S4) perfectly in ω, and the proof of the proposition is now complete. □

In Lemma 2.3.4 below, an integrability condition allows us to pass from discrete-time limits of the cocycle to continuous ones. This property is crucial to the proof of Theorem 2.2.1. The reason this property holds is because the integrability hypothesis together with the perfect ergodic theorem (Lemma 2.3.1 (ii)) allow for control of the excursions of the continuous-time cocycle between discrete times.

LEMMA 2.3.4. *Let $Y : \Omega \to H$ be a stationary point of the cocycle (U,θ) satisfying the integrability condition*

$$\int_\Omega \log^+ \sup_{0\le t_1,t_2\le a} \|U(t_2, Y(\theta(t_1,\omega)) + (\cdot), \theta(t_1,\omega))\|_{k,\epsilon}\, dP(\omega) < \infty$$

for any fixed $0<\rho, a<\infty$ and $\epsilon\in(0,1]$.

Define the random field $Z : \mathbf{R}^+ \times H \times \Omega \to H$ by

$$Z(t,x,\omega) := U(t, x+Y(\omega),\omega) - Y(\theta(t,\omega))$$

for $t\ge 0, x\in H, \omega\in\Omega$. Then (Z,θ) is a $C^{k,\epsilon}$ perfect cocycle. Furthermore, there is a sure event $\Omega_5 \in \mathcal{F}$ with the following properties:
(i) *$\theta(t,\cdot)(\Omega_5) = \Omega_5$ for all $t\in\mathbf{R}$,*
(ii) *For every $\omega\in\Omega_5$ and any $x\in H$, the statement*

$$\limsup_{n\to\infty} \frac{1}{n}\log|Z(n,x,\omega)| < 0 \tag{2.3.23}$$

implies

$$\limsup_{t\to\infty} \frac{1}{t}\log|Z(t,x,\omega)| = \limsup_{n\to\infty}\frac{1}{n}\log|Z(n,x,\omega)|. \tag{2.3.24}$$

PROOF. Note that, by definition, Z is a "centering" of the cocycle U with respect to the stationary trajectory $\{Y(\theta(t,\cdot)) : t \geq 0\}$ in the sense that $Z(t,0,\omega) = 0$ for all $(t,\omega) \in \mathbf{R}^+ \times \Omega$. Furthermore, (Z,θ) is a $C^{k,\epsilon}$ perfect cocycle. To see this let $t_1, t_2 \geq 0, \omega \in \Omega, x \in H$. Then by the perfect cocycle property for U, it follows that

$$\begin{aligned}
Z(t_2,&Z(t_1,x,\omega),\theta(t_1,\omega)) \\
&= U(t_2, Z(t_1,x,\omega) + Y(\theta(t_1,\omega)), \theta(t_1,\omega)) - Y(\theta(t_2,\theta(t_1,\omega))) \\
&= U(t_2, U(t_1, x + Y(\omega), \omega), \theta(t_1,\omega)) - Y(\theta(t_2 + t_1, \omega)) \\
&= Z(t_1 + t_2, x, \omega).
\end{aligned}$$

Using the integrability condition of the lemma, the proofs of assertions (i) and (ii) follow in the same manner as for the corresponding assertions in Lemma 3.4 ([M-S.2]). □

PROOF OF THEOREM 2.2.1. The proof of the theorem consists in two major undertakings:

(a) Using Ruelle's discrete-time analysis [Ru.2] to show that the assertions of Theorem 2.2.1 hold for the discretized cocycle, perfectly in ω.
(b) Extending the discrete-time results to continuous time via perfection techniques that are essentially based on the ergodic theorem and Kingman's subadditive ergodic theorem.

Recall the auxiliary cocycle (Z,θ) defined in Lemma 2.3.4. Consider the random family of maps $F_\omega : \bar{B}(0,1) \to H$, $\omega \in \Omega$, given by $F_\omega(x) := Z(1,x,\omega)$, $x \in H$, and the time-one shift $\tau := \theta(1,\cdot) : \Omega \to \Omega$. Adopting Ruelle's notation ([Ru.2], p. 272), we set $F_\omega^n := F_{\tau^{n-1}(\omega)} \circ \cdots \circ F_{\tau(\omega)} \circ F_\omega$. Therefore, $F_\omega^n = Z(n,\cdot,\omega)$ for each $n \geq 1$, because (Z,θ) is a cocycle. By Lemma 2.3.4, each map F_ω is $C^{k,\epsilon}$ ($\epsilon \in (0,1]$) on $\bar{B}(0,1)$ and by the definition of Z, it follows that $(DF_\omega)(0) = DU(1,Y(\omega),\omega)$. By the integrability hypothesis of the theorem, it is clear that $\log^+ \|DU(1,Y(\cdot),\cdot)\|_{L(H)}$ is integrable. Moreover, in view of the integrability hypothesis on (U,θ), it follows that the linearized continuous-time cocycle $(DU(t,Y(\omega),\omega),\theta(t,\omega))$ and the discrete-time cocycle $((DF_\omega^n)(0),\theta(n,\omega))$ share the same Lyapunov spectrum, viz.:

$$\{-\infty < \cdots < \lambda_{i+1} < \lambda_i < \cdots < \lambda_2 < \lambda_1\}.$$

(cf. [Mo.3]). Assume that λ_{i_0} is finite, that is $\lambda_{i_0} \in (-\infty, 0)$. Therefore, under hypotheses (I) of Theorem 5.1 in ([Ru.2], p. 272), there is a sure event $\Omega_1^* \in \mathcal{F}$ such that $\theta(t,\cdot)(\Omega_1^*) = \Omega_1^*$ for all $t \in \mathbf{R}$, $\bar{\mathcal{F}}$-measurable positive random variables $\rho_1, \beta_1 : \Omega_1^* \to (0,1)$, and a random family of $C^{k,\epsilon}$ ($k \geq 1, \epsilon \in (0,1]$) stable submanifolds $\tilde{S}_d(\omega)$ of $\bar{B}(0,\rho_1(\omega))$ satisfying the following properties for each $\omega \in \Omega_1^*$:
(2.3.25)
$$\tilde{S}_d(\omega) = \{x \in \bar{B}(0,\rho_1(\omega)) : |Z(n,x,\omega)| \leq \beta_1(\omega) e^{(\lambda_{i_0}+\epsilon_1)n} \text{ for all integers } n \geq 0\}.$$

When $\lambda_{i_0} = -\infty$, the stable manifold is defined by
(2.3.25′)
$$\tilde{S}_d(\omega) := \{x \in \bar{B}(0,\rho_1(\omega)) : |Z(n,x,\omega)| \leq \beta_1(\omega) e^{\lambda n} \text{ for all integers } n \geq 0\},$$

where $\lambda \in (-\infty, 0)$ is arbitrary. The stable subspace $\mathcal{S}(\omega)$ of the linearized cocycle $(DU(t,Y(\omega),\omega),\theta(t,\omega))$ is tangent to the stable manifold $\tilde{S}_d(\omega)$ at 0; viz. $T_0\tilde{S}_d(\omega) = \mathcal{S}(\omega)$. In particular, codim $\tilde{S}_d(\omega)$ is finite and non-random. Again by

Theorem 5.1 of [Ru.2]), we have the following estimate on the Lyapunov exponent of the Lipschitz constant of $(Z(n,\cdot),\theta(n,\cdot))$ over its stable manifold:

$$\limsup_{n\to\infty}\frac{1}{n}\log\left[\sup_{\substack{x_1,x_2\in\tilde{\mathcal{S}}_d(\omega)\\ x_1\neq x_2}}\frac{|Z(n,x_1,\omega)-Z(n,x_2,\omega)|}{|x_1-x_2|}\right]\leq\lambda_{i_0}. \tag{2.3.26}$$

The statements in the above paragraph hold for all ω in the $\theta(t,\cdot)$-invariant sure event Ω_1^*. In order to construct such an event, we will use perfection arguments and the proof of Theorem 5.1 ([Ru.2], p. 272). Assume first that $k=1$ and $\epsilon>0$. Using the notation of [Ru.2], denote $T^t(\omega):=DZ(t,0,\omega)$, $f(\omega):=\theta(1,\omega)$, $T_n(\omega):=DZ(1,0,\theta((n-1),\omega))$, for all $\omega\in\Omega$, any positive real t and any integer $n\geq 1$. It is possible to replace (5.3) in [Ru.2], p. 274) by its continuous-time *perfect* analogue

$$\lim_{t\to\infty}\frac{1}{t}\log^+\|Z(1,\cdot,\theta(t,\omega))\|_{1,\epsilon}=0. \tag{2.3.27}$$

This follows from the integrability hypothesis of the theorem and the perfect ergodic theorem (Lemma 2.3.1 (ii)). More specifically, (2.3.27) holds for all ω in a sure event $\Omega_1^*\in\mathcal{F}$ such that $\theta(t,\cdot)(\Omega_1^*)=\Omega_1^*$ for all $t\in\mathbf{R}$. Assume $\lambda_{i_0}>-\infty$. Adopting the terminology of Theorem 1.1 ([Ru.2], p. 248), take $S=i_0-1$, and $\mu^{(S+1)}=\lambda_{i_0}$. In case $\lambda_{i_0}=-\infty$, take $\mu^{(S+1)}$ to be any fixed number in $(-\infty,0)$. The integrability hypothesis on U coupled with Lemma 2.3.3 (where $j_0=i_0-1$) imply the existence of a sure event $\Omega_2^*\in\mathcal{F}$ such that $\Omega_2^*\subseteq\Omega_1^*$, $\theta(t,\cdot)(\Omega_2^*)=\Omega_2^*$ for all $t\in\mathbf{R}$, and the sequence $T_n(\omega), V_n(\omega):=E_{i_0}(\tau^n(\omega))$, $n\geq 1$, satisfies Conditions (S) of [Ru.2], p. 256) for every $\omega\in\Omega_2^*$. Pick and fix any $\omega\in\Omega_2^*$. As in the proof of Theorem 5.1 ([Ru.2], pp. 274-278), the "perturbation theorem" (Theorem 4.1, [Ru.2], pp. 262-263) holds for the sequence $T_n(\omega), n\geq 1$. Thus the assertions in the previous paragraph are valid for $k=1$ and any $\epsilon\in(0,1]$. When $k>1$ and $\epsilon\in(0,1]$), we first apply the previous analysis to the perfect cocycle

$$\left(\check{Z}(t,x,x_1,\omega):=(Z(t,x,\omega),DZ(t,x,\omega)x_1),\theta(t,\omega)\right),\quad x,x_1\in H,\ t\geq 0,$$

on $H\oplus H$. Secondly, we use the inductive argument of ([Ru.2], pp. 278-279) to show that the $\tilde{\mathcal{S}}_d(\omega)$ are $C^{k,\epsilon}$ manifolds ($k>1,\epsilon\in(0,1]$) *perfectly in* ω.

To establish assertion (a) of the theorem, let $\tilde{\mathcal{S}}(\omega),\omega\in\Omega_1^*$, be the set defined therein. Then the definition of Z and property (2.3.25) of $\tilde{\mathcal{S}}_d(\omega)$ imply that

$$\tilde{\mathcal{S}}(\omega)=\tilde{\mathcal{S}}_d(\omega)+Y(\omega) \tag{2.3.28}$$

for all $\omega\in\Omega_1^*$. Thus $\tilde{\mathcal{S}}(\omega)$ is a $C^{k,\epsilon}$ manifold ($k>1,\epsilon\in(0,1]$), with tangent space $T_{Y(\omega)}\tilde{\mathcal{S}}(\omega)=T_0\tilde{\mathcal{S}}_d(\omega)=\mathcal{S}(\omega)$. In particular, codim $\tilde{\mathcal{S}}(\omega)=$ codim $\mathcal{S}(\omega),\omega\in\Omega_1^*$, is finite and non-random.

To complete the proof of the inequality (2.2.1) in part (a) of the theorem, use (2.3.26) to get

$$\limsup_{n\to\infty}\frac{1}{n}\log|Z(n,x,\omega)|\leq\lambda_{i_0}$$

perfectly in ω for all $x\in\tilde{\mathcal{S}}_d(\omega)$. In view of Lemma 2.3.4, we may extend the above estimate to cover its continuous-time counterpart. Hence we obtain a sure event

$\Omega_3^* \subseteq \Omega_2^*$, $\Omega_3^* \in \mathcal{F}$, such that $\theta(t,\cdot)(\Omega_3^*) = \Omega_3^*$ for all $t \in \mathbf{R}$, and

$$\limsup_{t\to\infty} \frac{1}{t} \log |Z(t,x,\omega)| \leq \lambda_{i_0} \tag{2.3.29}$$

for all $\omega \in \Omega_3^*$ and all $x \in \tilde{\mathcal{S}}_d(\omega)$. The above inequality together with definition of Z imply the estimate (2.2.1) of the theorem.

Next, we establish assertion (b) of the theorem. To do so, let $\omega \in \Omega_1^*$ and $x \in \tilde{\mathcal{S}}_d(\omega)$. Then by (2.3.26), it follows that there is a positive integer $N_0 := N_0(\omega)$, independent of $x \in \tilde{\mathcal{S}}_d(\omega)$, such that $Z(n,x,\omega) \in \bar{B}(0,1)$ for all $n \geq N_0$. Now Lemma 2.3.1(ii) gives a $\theta(t,\cdot)$-invariant sure event Ω_3 such that

$$\lim_{t\to\infty} \frac{1}{t} \log^+ \sup_{\substack{0\leq u\leq 1,\\ (v^*,\eta^*)\in\bar{B}(0,1)}} \|DZ(u,(v^*,\eta^*),\theta(t,\omega))\|_{L(H)} = 0 \tag{2.3.29$'$}$$

for all $\omega \in \Omega_3$. Define the sure event $\Omega_4^* := \Omega_3^* \cap \Omega_3 \in \mathcal{F}$. Clearly, $\theta(t,\cdot)(\Omega_4^*) = \Omega_4^*$ for all $t \in \mathbf{R}$. By the definition of Z and the Mean Value Theorem, we obtain the following inequalities

$$\begin{aligned}
\sup_{n\leq t\leq n+1} \frac{1}{t} \log&\left[\sup_{\substack{x_1\neq x_2,\\ (v_1,\eta_1),x_2\in\tilde{\mathcal{S}}(\omega)}} \frac{|U(t,x_1,\omega)-U(t,x_2,\omega)|}{|x_1-x_2|}\right] \\
&= \sup_{n\leq t\leq n+1} \frac{1}{t} \log\left[\sup_{\substack{x_1\neq x_2,\\ x_1,x_2\in\tilde{\mathcal{S}}_d(\omega)}} \frac{|Z(t,x_1,\omega)-Z(t,x_2,\omega)|}{|x_1-x_2|}\right] \\
&\leq \frac{1}{n} \log^+ \sup_{\substack{0\leq u\leq 1,\\ (v^*,\eta^*)\in\bar{B}(0,1)}} \|DZ(u,(v^*,\eta^*),\theta(n,\omega))\|_{L(H)} \\
&\qquad + \frac{n}{(n+1)} \frac{1}{n} \log\left[\sup_{\substack{x_1\neq x_2,\\ x_1,x_2\in\tilde{\mathcal{S}}_d(\omega)}} \frac{|Z(n,x_1,\omega)-Z(n,x_2,\omega)|}{|x_1-x_2|}\right]
\end{aligned}$$

for all $\omega \in \Omega_4^*$, and all sufficiently large $n \geq N_0(\omega)$. Now take $\limsup_{n\to\infty}$ on both sides of the above inequality, and use (2.3.26), (2.3.29$'$) in order to complete the proof assertion (b) of the theorem.

The cocycle invariance (2.2.3) in part (c) of the theorem follows immediately from the Oseledec-Ruelle theorem (Theorem 2.1.1) applied to the perfect linearized cocycle $(DU(t,Y(\omega),\omega),\theta(t,\omega))$. Indeed, one gets a sure $\theta(t,\cdot)$-invariant event $\in \mathcal{F}$ (also denoted by Ω_1^*), such that $DU(t,Y(\omega),\omega)(\mathcal{S}(\omega)) \subseteq \mathcal{S}(\theta(t,\omega))$ for all $t \geq 0$ and all $\omega \in \Omega_1^*$.

The proof of the asymptotic invariance property (2.2.2) of the non-linear cocycle requires some work. To achieve this, we will extend the arguments underlying the proofs of Theorems 5.1 and 4.1 in [Ru.2], pp. 262-279, to a continuous time setting. The crucial step towards this goal is to show that the two random variables ρ_1, β_1 in (2.3.25) may be redefined on a sure event (also denoted by) Ω_1^* such that $\theta(t,\cdot)(\Omega_1^*) = \Omega_1^*$ for all $t \in \mathbf{R}$, and

$$\rho_1(\theta(t,\omega)) \geq \rho_1(\omega)e^{(\lambda_{i_0}+\epsilon_1)t}, \quad \beta_1(\theta(t,\omega)) \geq \beta_1(\omega)e^{(\lambda_{i_0}+\epsilon_1)t} \tag{2.3.30}$$

for every $\omega \in \Omega_1^*$ and all $t \geq 0$. For the given choice of ϵ_1, fix $0 < \epsilon_3 < -\epsilon(\lambda_{i_0}+\epsilon_1)/4$, where $\epsilon \in (0,1]$ denotes the Hölder exponent of U. The above inequalities hold in the *discrete* case (when $t = n$, a positive integer) because of Theorem 5.1 (c) ([Ru.2],

p. 274). To prove them for any continuous time t, we will modify the definitions of ρ_1, β_1 in the proofs of Theorems 5.1 and 4.1 in [Ru.2]. In the notation of the proof of Theorem 5.1 ([Ru.2], p. 274), we replace the random variable G in (5.4) ([Ru.2], p. 274) by the larger one

$$\tilde{G}(\omega) := \sup_{t\geq 0} \|Z(1,\cdot,\theta(t,\omega))\|_{1,\epsilon}\, e^{(-t\epsilon_3-\lambda\epsilon)}. \tag{2.3.31}$$

Clearly, $\tilde{G}(\omega)$ is finite perfectly in ω, because of (2.3.27) and Lemma 2.3.2. Following ([Ru.2], pp. 266, 274), the random variables ρ_1, β_1 may be chosen according to the relations

$$\beta_1 := \left[\frac{\delta_1 \wedge \left(\frac{1}{\sqrt{2}A}\right)}{2\tilde{G}}\right]^{\frac{1}{\epsilon}} \wedge 1 \tag{2.3.32}$$

$$\rho_1 := \frac{\beta_1}{B_{\epsilon_3}} \tag{2.3.33}$$

where A, δ_1 and B_{ϵ_3} are random positive constants that are defined via continuous-time analogues of the relations (4.26), (4.18)-(4.21), (4.24), (4.25) in [Ru.2], pp. 265-267, with η replaced by ϵ_3. In particular, the "ancestry" of A, δ_1 and B_{ϵ_3} in Ruelle's argument may be traced back to the constants $D_{\epsilon_3}, K_{\epsilon_3}$ which appear in Lemmas 2.3.3 and 2.3.2 of this article. Hence (2.3.30) will follow if we can show that, for sufficiently small $\epsilon_3 > 0$, the following inequalities

$$\left.\begin{aligned} K_{\epsilon_3}(\theta(l,\omega)) &\leq K_{\epsilon_3}(\omega) + \frac{\epsilon_3 l}{2} \\ D_{\epsilon_3}(\theta(l,\omega)) &\leq e^{\frac{\epsilon_3 l}{2}} D_{\epsilon_3}(\omega) \\ \tilde{G}(\theta(l,\omega)) &\leq e^{\epsilon_3 l}\tilde{G}(\omega) \end{aligned}\right\} \tag{2.3.34}$$

hold perfectly in ω for all real $l \geq 0$. The first and second inequalities in (2.3.34) follow from Lemmas 2.3.2(ii) and 2.3.3, respectively. The third inequality is an immediate consequence of the definition of $\tilde{G}$ in (2.3.31). The proof of (2.3.30) is now complete in view of (2.3.32), (2.3.33) and (2.3.34).

The inequalities in (2.3.30) will allow us to establish the asymptotic invariance property (2.2.2) in (c) of the theorem. By the perfect inequality in (b), there is a sure event $\Omega_5^* \subseteq \Omega_4^*$ such that $\theta(t,\cdot)(\Omega_5^*) = \Omega_5^*$ for all $t \in \mathbf{R}$, and for any $0 < \epsilon' < \epsilon_1$ and any $\omega \in \Omega_5^*$, there exists $\beta^{\epsilon'}(\omega) > 0$ (independent of x) so that

$$|U(t,x,\omega) - Y(\theta(t,\omega))| \leq \beta^{\epsilon'}(\omega)e^{(\lambda_{i_0}+\epsilon')t} \tag{2.3.35}$$

for all $x \in \tilde{\mathcal{S}}(\omega)$, $t \geq 0$. Let t be any positive real, n a non-negative integer, $\omega \in \Omega_5^*$ and $x \in \tilde{\mathcal{S}}(\omega)$. Using the cocycle property and (2.3.35), we obtain

$$\begin{aligned} |U(n,U(t,x,\omega),\theta(t,\omega)) - Y(\theta(n,\theta(t,\omega)))| &= |U(n+t,x,\omega) - Y(\theta(n+t,\omega))| \\ &\leq \beta^{\epsilon'}(\omega)e^{(\lambda_{i_0}+\epsilon')(n+t)} \\ &\leq \beta^{\epsilon'}(\omega)e^{(\lambda_{i_0}+\epsilon')t}e^{(\lambda_{i_0}+\epsilon_1)n}. \end{aligned} \tag{2.3.36}$$

Using (2.3.30), (2.3.35), (2.3.36) and the definition of $\tilde{\mathcal{S}}(\theta(t,\omega))$, we see that for each $\omega \in \Omega_5^*$, there exists $\tau_1(\omega) > 0$ such that $U(t,x,\omega) \in \tilde{\mathcal{S}}(\theta(t,\omega))$ for all $t \geq \tau_1(\omega)$. Hence, for all $\omega \in \Omega_5^*$,

$$U(t,\cdot,\omega)(\tilde{\mathcal{S}}(\omega)) \subseteq \tilde{\mathcal{S}}(\theta(t,\omega)), \quad t \geq \tau_1(\omega)$$

and the proof of assertion (c) is complete.

Our next objective is to establish the existence of the *perfect* family of local unstable manifolds $\tilde{\mathcal{U}}(\omega)$ in assertion (d) of the theorem. To this end, we define the random field $\hat{Z} : \mathbf{R}^+ \times H \times \Omega \to H$ by

$$\hat{Z}(t, x, \omega) := U(t, x + Y(\theta(-t, \omega)), \theta(-t, \omega)) - Y(\omega) \tag{2.3.37}$$

for all $t \geq 0$, $x \in H$, $\omega \in \Omega$. Note that $\hat{Z}(t, \cdot, \omega) = Z(t, \cdot, \theta(-t, \omega))$, $t \geq 0$, $\omega \in \Omega$; and $\hat{Z}$ is $(\mathcal{B}(\mathbf{R}^+) \otimes \mathcal{B}(H) \otimes \mathcal{F}, \mathcal{B}(H))$-measurable. Since Y is a stationary point for (U, θ), we may replace ω by $\theta(-t, \omega)$ in (2.1.1). Thus $\hat{Z}(t, 0, \omega) = 0$ for all $t \geq 0$, $\omega \in \Omega$. We contend that $([D\hat{Z}(t, 0, \omega)]^*, \theta(-t, \omega), t \geq 0)$ is a perfect linear cocycle in $L(H)$. To see this, we first observe that $(DU(t, Y(\omega), \omega), \theta(t, \omega))$ is an $L(H)$-valued perfect cocycle:

$$DU(t_1 + t_2, Y(\omega), \omega) = DU(t_1, Y(\theta(t_2, \omega)), \theta(t_2, \omega)) \circ DU(t_2, Y(\omega), \omega)$$

for all $\omega \in \Omega, t_1, t_2 \geq 0$. Secondly, we replace ω by $\theta(-t_1 - t_2, \omega)$ and take adjoints in the above identity to obtain

$$\begin{aligned} &[DU(t_1 + t_2, Y(\theta(-t_1 - t_2, \omega)), \theta(-t_1 - t_2, \omega))]^* \\ &= [DU(t_2, Y(\theta(-t_1 - t_2, \omega)), \theta(-t_1 - t_2, \omega))]^* \circ [DU(t_1, Y(\theta(-t_1, \omega)), \theta(-t_1, \omega)]^* \end{aligned}$$

for all $\omega \in \Omega, t_1, t_2 \geq 0$. Therefore,

$$[D\hat{Z}(t_1 + t_2, 0, \omega)]^* = [D\hat{Z}(t_2, 0, \theta(-t_1, \omega))]^* \circ [D\hat{Z}(t_1, 0, \omega)]^*$$

for all $\omega \in \Omega, t_1, t_2 \geq 0$; and our contention is proved.

We will now apply the Oseledec-Ruelle theorem to the perfect cocycle $([D\hat{Z}(t, 0, \omega)]^*, \theta(-t, \omega), t \geq 0)$. To do this, it is sufficient to check the integrability condition

$$\int_\Omega \log^+ \sup_{0 \leq t_1, t_2 \leq a} \|[D\hat{Z}(t_2, 0, \theta(-t_1, \omega))]^*\|_{L(H)} \, dP(\omega) < \infty \tag{2.3.38}$$

for any fixed $a \in (0, \infty)$. The above integrability relation follows from the integrability hypothesis of Theorem 2.2.1 and the P-preserving property of $\theta(t, \cdot)$:

$$\begin{aligned} &\int_\Omega \log^+ \sup_{0 \leq t_1, t_2 \leq a} \|[D\hat{Z}(t_2, 0, \theta(-t_1, \omega))]^*\|_{L(H)} \, dP(\omega) \\ &= \int_\Omega \log^+ \sup_{0 \leq t_1, t_2 \leq a} \|DU(t_2, Y(\theta(-t_2 - t_1, \omega)), \theta(-t_2 - t_1, \omega)))\|_{L(H)} \, dP(\omega) \\ &\leq \int_\Omega \log^+ \sup_{0 \leq t_1 \leq 2a, \, 0 \leq t_2 \leq a} \|DU(t_2, Y(\theta(t_1, \omega)), \theta(t_1, \omega)))\|_{L(H)} \, dP(\omega) \\ &\leq \int_\Omega \log^+ \sup_{0 \leq t_1 \leq a, \, 0 \leq t_2 \leq a} \|DU(t_2, Y(\theta(t_1, \omega)), \theta(t_1, \omega)))\|_{L(H)} \, dP(\omega) \\ &+ \int_\Omega \log^+ \sup_{a \leq t_1 \leq 2a, \, 0 \leq t_2 \leq a} \|DU(t_2, Y(\theta(t_1 - a, \omega)), \theta(t_1 - a, \omega)))\|_{L(H)} \, dP(\omega) \\ &= 2 \int_\Omega \log^+ \sup_{0 \leq t_1, t_2 \leq a} \|DU(t_2, Y(\theta(t_1, \omega)), \theta(t_1, \omega)))\|_{L(H)} \, dP(\omega) < \infty. \end{aligned}$$

By (2.3.38) and the Oseledec-Ruelle theorem, we conclude that the linear cocycle $([D\hat{Z}(t,0,\omega)]^*, \theta(-t,\omega), t \geq 0)$ has a fixed discrete Lyapunov spectrum. Furthermore, this spectrum (with multiplicities) coincides with that of the cocycle $(DU(t,Y(\omega),\omega), \theta(t,\omega))$, viz. $\{\cdots \lambda_{i+1} < \lambda_i < \cdots < \lambda_2 < \lambda_1\}$ where $\lambda_i \neq 0$ for all $i \geq 1$, by hyperbolicity. See [Ru.2], Section 3.5, p. 261.

The next step in our construction of the *perfect* random family of local unstable manifolds $\tilde{\mathcal{U}}(\omega)$ starts with the following estimate:

$$\int_\Omega \log^+ \sup_{0 \leq t_1, t_2 \leq 1} \|\hat{Z}(t_2, \cdot, \theta(-t_1,\omega))\|_{k,\epsilon}\, dP(\omega) < \infty.$$

By the same argument as in the previous paragraph, the above estimate is a consequence of the P-preserving property of $\theta(t,\cdot), t \in \mathbf{R}$, and the integrability hypothesis of the theorem. Define λ_{i_0-1} as in the statement of Theorem 2.2.1, and fix any $\epsilon_2 \in (0, \lambda_{i_0-1})$. In view of the above integrability property, it follows from Lemma 2.3.3 that the sequence $\tilde{T}_n(\omega) := [D\hat{Z}(1,0,\theta(-n,\omega))]^*$, $\theta(-n,\omega)$, $n \geq 0$, satisfies Condition (S) of [Ru.2] perfectly in ω. Hence the sequence $\tilde{T}_n(\omega), n \geq 1$, satisfies Corollary 3.4 ([Ru.2], p. 260) perfectly in ω, because of Proposition 3.3 in [Ru.2]. At this point, we may modify the arguments in the proof of Ruelle's Theorem 6.1 ([Ru.2], p. 280) using an approach analogous to the one used in constructing the stable manifolds in this proof. Therefore, one gets a $\theta(-t,\cdot)$-invariant sure event $\hat{\Omega}_1^* \in \mathcal{F}$ and $\bar{\mathcal{F}}$-measurable random variables $\rho_2, \beta_2 : \hat{\Omega}_1^* \to (0,1)$ satisfying the following properties. If $\lambda_{i_0-1} < \infty$, define $\tilde{\mathcal{U}}_d(\omega)$ to be the set of all $x_0 \in \bar{B}(0,\rho_2(\omega))$ with the property that there is a discrete "history" process $u(-n,\cdot) : \Omega \to H, n \geq 0$, such that $u(0,\omega) = x_0$, $\hat{Z}(1, u(-(n+1),\omega), \theta(-n,\omega)) = u(-n,\omega)$ and $|u(-n,\omega)| \leq \beta_2(\omega)e^{-n(\lambda_{i_0-1}-\epsilon_2)}$ for all $n \geq 0$. If $\lambda_{i_0-1} = \infty$, define $\tilde{\mathcal{U}}_d(\omega)$ to be the set of all $x_0 \in H$ with the property that there is a discrete history process $u(-n,\cdot) : \Omega \to H, n \geq 0$, such that $u(0,\omega) = x_0$, and $|u(-n,\omega)| \leq \beta_2(\omega)e^{-\lambda n}$ for all $n \geq 0$ and arbitrary $\lambda > 0$. It follows from ([Ru.2], p. 281) that the discrete history process $u(-n,\cdot)$ is uniquely determined by x_0. Moreover, each $\tilde{\mathcal{U}}_d(\omega), \omega \in \hat{\Omega}_1^*$, is a $C^{k,\epsilon}$ $(k \geq 1, \epsilon \in (0,1])$ finite-dimensional submanifold of $\bar{B}(0,\rho_2(\omega))$ with tangent space $\mathcal{U}(\omega)$ at 0, and $\dim \tilde{\mathcal{U}}_d(\omega)$ is fixed independently of ω and ϵ_2. Furthermore,

$$\rho_2(\theta(-t,\omega)) \geq \rho_2(\omega)e^{-(\lambda_{i_0-1}-\epsilon_2)t}, \quad \beta_2(\theta(-t,\omega)) \geq \beta_2(\omega)e^{-(\lambda_{i_0-1}-\epsilon_2)t}. \tag{2.3.39}$$

perfectly in ω for all $t \geq 0$. We claim that the set $\tilde{\mathcal{U}}(\omega)$ defined in (d) of Theorem 2.2.1 coincides with $\tilde{\mathcal{U}}_d(\omega) + Y(\omega)$ for each $\omega \in \hat{\Omega}_1^*$. We first show that $\tilde{\mathcal{U}}_d(\omega) + Y(\omega) \subseteq \tilde{\mathcal{U}}(\omega)$. Let $x_0 \in \tilde{\mathcal{U}}_d(\omega)$ and u be as above. Set

$$y_0(-n,\omega) := u(-n,\omega) + Y(\theta(-n,\omega)), \qquad n \geq 0. \tag{2.3.40}$$

It is easy to check that y_0 is a discrete history process satisfying the first and second assertions in (d) of the theorem. Hence $x_0 + Y(\omega) \in \tilde{\mathcal{U}}(\omega)$. Similarly, $\tilde{\mathcal{U}}(\omega) \subseteq \tilde{\mathcal{U}}_d(\omega) + Y(\omega)$ for all $\omega \in \hat{\Omega}_1^*$. Hence $\tilde{\mathcal{U}}(\omega) = \tilde{\mathcal{U}}_d(\omega) + Y(\omega)$ for all $\omega \in \hat{\Omega}_1^*$. This immediately implies that $\tilde{\mathcal{U}}(\omega)$ is a $C^{k,\epsilon}$ $(k \geq 1, \epsilon \in (0,1])$ finite-dimensional submanifold of $\bar{B}(Y(\omega),\rho_2(\omega))$, and

$$T_{Y(\omega)}\tilde{\mathcal{U}}(\omega) = T_0\tilde{\mathcal{U}}_d(\omega) = \mathcal{U}(\omega)$$

for all $\omega \in \hat{\Omega}_1^*$.

We will next address the issue of the existence of the continuous-time history process satisfying the third assertion in part (d) of the theorem. Suppose $x \in \tilde{\mathcal{U}}(\omega)$. From what we proved in the previous paragraph, it follows that there is an $x_0 \in \mathcal{U}_d(\omega)$ such that $x = x_0 + Y(\omega)$. The discrete process y_0 given by (2.3.40) may be extended to a continuous-time history process $y(\cdot,\omega) : (-\infty, 0] \to H$ such that $y(0,\omega) = x$, and $y(\cdot,\omega)$ satisfies the third assertion in (d). This is achieved by interpolation within the periods $[-(n+1), -n]$, $n \geq 0$, using the cocycle property of U: Indeed, let $s \in (-(n+1), -n)$. Then there is an $\alpha \in (0,1)$, such that $s = \alpha - (n+1)$. Define

$$y(s,\omega) := U(s+n+1, y_0(-(n+1),\omega), \theta(-(n+1),\omega)).$$

Obviously, $y(0,\omega) = x_0 + Y(\omega) = x$. Let $s \in (-(n+1), -n)$ and suppose $0 < t \leq -s$. Pick a positive integer $m < n$ such that $s+t \in [-(m+1), -m]$. The above definition of y, together with the perfect cocycle property for U, easily imply that

$$y(t+s,\omega) = U(t, y(s,\omega), \theta(s,\omega)). \tag{2.3.41}$$

In particular, $U(t, y(-t,\omega), \theta(-t,\omega)) = x$ for all $t \geq 0$. This follows from (2.3.41) when s is replaced by $-t$. Furthermore, for each $x \in \tilde{\mathcal{U}}(\omega)$, the above continuous-time history process is uniquely determined because its discrete-time counterpart is unique.

We will now prove the following estimate

$$\limsup_{t\to\infty} \frac{1}{t} \log |y(-t,\omega) - Y(\theta(-t,\omega))| \leq -\lambda_{i_0-1} \tag{2.3.42}$$

perfectly in ω. We start with its discrete-time counterpart

$$\limsup_{n\to\infty} \frac{1}{n} \log |y(-n,\omega) - Y(\theta(-n,\omega))| \leq -\lambda_{i_0-1} \tag{2.3.43}$$

which holds perfectly in ω, because of Theorem 6.1 (b) in [Ru.2]. Let $t \in (n, n+1)$. Then there exists $\gamma \in (0,1)$ such that $-t = \gamma - (n+1)$. Thus, by the definition of y and the Mean Value Theorem, it follows that

$$\begin{aligned}
&|y(-t,\omega) - Y(\theta(-t,\omega))| \\
&= |U(\gamma, y(-(n+1),\omega), \theta(-(n+1),\omega)) - U(\gamma, Y(\theta(-(n+1),\omega), \theta(-(n+1),\omega))| \\
&\leq \sup_{\substack{(v^*,\eta^*)\in \bar{B}(0,1),\\ \gamma\in(0,1)}} \|DU(\gamma, (v^*,\eta^*) + Y(\theta(-(n+1),\omega)), \theta(-(n+1),\omega))\|_{L(H)} \\
&\qquad \times |y(-(n+1),\omega) - Y(\theta(-(n+1),\omega)))|
\end{aligned}$$

perfectly in ω. Hence,

$$\begin{aligned}
&\limsup_{t\to\infty} \frac{1}{t} \log |y(-t,\omega) - Y(\theta(-t,\omega))| \\
&\quad \leq \limsup_{n\to\infty} \frac{1}{n} \log^+ \sup_{\substack{(v^*,\eta^*)\in \bar{B}(0,1),\\ \gamma\in(0,1)}} \|DU(\gamma, (v^*,\eta^*) + \\
&\qquad\qquad + Y(\theta(-(n+1),\omega)), \theta(-(n+1),\omega))\|_{L(H)} \\
&\quad + \limsup_{n\to\infty} \frac{1}{n} \log |y(-(n+1),\omega) - Y(\theta(-(n+1),\omega)))|.
\end{aligned}$$

By the integrability condition of the theorem and the perfect ergodic theorem (Lemma 2.3.1 (ii)), the first term on the right hand side of the above inequality is zero, perfectly in $\omega \in \Omega$. Since $y(0) \in \tilde{\mathcal{U}}(\omega)$, the second term is less than or equal to $-\lambda_{i_0-1}$. This completes the proof of assertion (d) of the theorem.

We will omit the proof of assertion (e), since it is very similar to that of (2.3.42).

Our next objective is to prove assertion (f) of the theorem. Note first that the perfect invariance

$$DU(t,\cdot,\theta(-t,\omega))(\mathcal{U}(\theta(-t,\omega))) = \mathcal{U}(\omega), \quad t \geq 0,$$

follows from the cocycle property for the linearized semiflow and Theorem 2.1.2; cf. [Mo.3], Corollary 2 (v) of Theorem 4. Since $\dim \mathcal{U}(\omega)$ is fixed and finite perfectly in ω, the restriction

$$DU(t,\cdot,\theta(-t,\omega))|\mathcal{U}(\theta(-t,\omega)) : \mathcal{U}(\theta(-t,\omega)) \to \mathcal{U}(\omega), \quad t \geq 0,$$

is a linear homeomorphism onto. It remains to check the following asymptotic invariance property in (f):

$$\tilde{\mathcal{U}}(\omega) \subseteq U(t,\cdot,\theta(-t,\omega))(\tilde{\mathcal{U}}(\theta(-t,\omega))), \quad t \geq \tau_2(\omega), \tag{2.3.44}$$

perfectly in ω for some $\tau_2(\omega) > 0$. Suppose $x \in \tilde{\mathcal{U}}(\omega)$. Then by assertions (d), (e) of the theorem and inequalities (2.3.39), there exist a (unique) history process $y(-t,\omega), t \geq 0$, and a random time $\tau_2(\omega) > 0$ satisfying the following: $y(0,\omega) = x$, $y(-t,\omega) \in \bar{B}(Y(\theta(-t,\omega)), \rho_2(\theta(-t,\omega)))$ for all $t \geq \tau_2(\omega)$, and

$$y(t'-t,\omega) = U(t', y(-t,\omega), \theta(-t,\omega)), \quad 0 < t' \leq t, \tag{2.3.45}$$

perfectly in ω. Pick any $t_1 \geq \tau_2(\omega)$. Then $x = U(t_1, y(-t_1,\omega), \theta(-t_1,\omega))$, because of (2.3.45) (for $t = t' = t_1$). Now $y(-t_1,\omega) \in \tilde{\mathcal{U}}(\theta(-t_1,\omega)))$. To prove this, we define the process $y_1(-t,\omega) := y(-t-t_1,\omega)$, $t \geq 0$. Hence $y_1(\cdot,\omega)$ is a history process and

$$y_1(0,\omega) = y(-t_1,\omega) \in \bar{B}(Y(\theta(-t_1,\omega)), \rho_2(\theta(-t_1,\omega))).$$

Therefore $y(-t_1,\omega) \in \tilde{\mathcal{U}}(\theta(-t_1,\omega)))$. This implies (2.3.44) because $t_1 \geq \tau_2(\omega)$ is arbitrary.

To prove the transversality property in in (g), note the following perfect identities:

$$T_{Y(\omega)}\tilde{\mathcal{U}}(\omega) = \mathcal{U}(\omega), \quad T_{Y(\omega)}\tilde{\mathcal{S}}(\omega) = \mathcal{S}(\omega), \quad H = \mathcal{U}(\omega) \oplus \mathcal{S}(\omega).$$

All the assertions (a)-(g) of the theorem will hold perfectly in ω if we take $\Omega^* := \Omega_1^* \cap \hat{\Omega}_1^*$.

To deal with the case when U is a C^∞ cocycle, we adapt the proof in [Ru.2], section (5.3) (p. 297). Thus we obtain a $\theta(t,\cdot)$-invariant sure event in $\mathcal{F}$ (also denoted by Ω^*) such that $\tilde{\mathcal{S}}(\omega)$ and $\tilde{\mathcal{U}}(\omega)$ are C^∞ for all $\omega \in \Omega^*$. This completes the proof of Theorem 2.2.1. □

2.4. The local stable manifold theorem for see's and spde's

In this section, we discuss several classes of semilinear stochastic evolutions equations and spde's. The objective is to establish sufficient conditions for a local stable manifold theorem for each class.

(a) Semilinear see's: Additive noise.

Let K, H be two separable real Hilbert spaces. Let A be a self-adjoint operator on H such that $A \geq cI_H$, where c is a real constant and I_H is the identity operator on H. Assume that A admits a discrete non-vanishing spectrum $\{\mu_n, n \geq 1\}$ which is bounded below. Let $\{e_n, n \geq 1\}$ denote a basis for H consisting of eigen vectors of A, viz. $Ae_n = \mu_n e_n$, $n \geq 1$. Assume further that A^{-1} is trace-class. Suppose $B_0 \in L_2(K, H)$. Let $W(t), t \in \mathbf{R}$, be a Brownian motion on the canonical complete filtered Wiener space $(\Omega, \bar{\mathcal{F}}, (\mathcal{F}_t)_{t\geq 0}, P)$ and with covariance Hilbert space K (Section 1.2). Let $T_t = e^{-At}$ stand for the strongly continuous semigroup generated by $-A$.

Denote by μ_m the largest negative eigenvalue of A and by μ_{m+1} its smallest positive eigenvalue. Thus there is an orthogonal $\{T_t\}_{t\geq 0}$-invariant splitting of H using the negative eigenvalues $\{\mu_1, \mu_2, \cdots, \mu_m\}$ and the positive eigenvalues $\{\mu_n : n \geq m+1\}$ of A:

$$H = H^+ \oplus H^-$$

where H^+ is a closed linear subspace of H and H^- is a finite-dimensional subspace. Denote by $p^+ : H \to H^+$ and $p^- : H \to H^-$ the corresponding projections onto H^+ and H^- respectively. Since H^- is finite-dimensional, then $T_t|H^-$ is invertible for each $t \geq 0$. Therefore, we can set $T_{-t} := [T_t|H^-]^{-1} : H^- \to H^-$ for each $t \geq 0$.

Consider the following semilinear see on H:

$$\begin{aligned} du(t) &= -Au(t)\,dt + F(u(t))\,dt + B_0 dW(t), \quad t \geq 0, \\ u(0) &= x \in H. \end{aligned} \tag{2.4.1}$$

In the above equation, let $F : H \to H$ be a globally Lipschitz map with Lipschitz constant L:

$$|F(v_1) - F(v_2)| \leq L|v_1 - v_2|, \quad v_1, v_2 \in H.$$

Then (2.4.1) has a unique mild solution given by

$$u(t, x) = T_t x + \int_0^t T_{t-s} F(u(s, x))ds + \int_0^t T_{t-s} B_0 dW(s), \quad t \geq 0 \tag{2.4.2}$$

Furthermore, if $F : H \to H$ is $C^{k,\epsilon}$, the mild solution of (2.4.2) generates a $C^{k,\epsilon}$ perfect cocycle also denoted by $u : \mathbf{R}^+ \times H \times \Omega \to H$.

Suppose that $F : H \to H$ is globally bounded, and its Lipschitz constant L satisfies

$$L[\mu_{m+1}^{-1} - \mu_m^{-1}] < 1. \tag{2.4.3}$$

Note that the above condition is automatically satisfied in the affine linear case $F \equiv 0$.

The next proposition is key to the existence and uniqueness of a stationary random point for the cocycle (u, θ) in the sense of Definition 2.1.1.

PROPOSITION 2.4.1. *Assume the above conditions on A, B_0, F together with (2.4.3). Then there is a unique $\mathcal{F}$-measurable map $Y : \Omega \to H$ satisfying*

$$
\begin{aligned}
Y(\omega) = &\int_{-\infty}^{0} T_{-s}p^+ F(Y(\theta(s,\omega)))ds - \int_{0}^{\infty} T_{-s}p^- F(Y(\theta(s,\omega)))ds \\
&+\left[\int_{-\infty}^{0} T_{-s}p^+ B_0\, dW(s)\right](\omega) - \left[\int_{0}^{\infty} T_{-s}p^- B_0\, dW(s)\right](\omega)
\end{aligned}
\tag{2.4.4}
$$

for all $\omega \in \Omega$.

PROOF. We use a contraction mapping argument to show that the integral equation (2.4.4) has an $\mathcal{F}$-measurable solution $Y : \Omega \to H$.

Define the $\mathcal{F}$-measurable map $Y_1 : \Omega \to H$ by

$$
Y_1(\omega) := \left[\int_{-\infty}^{0} T_{-s}p^+ B_0\, dW(s)\right](\omega) - \left[\int_{0}^{\infty} T_{-s}p^- B_0\, dW(s)\right](\omega), \quad \omega \in \Omega.
$$

Denote by $B(\Omega, H)$ the Banach space of all (surely) bounded $\mathcal{F}$-measurable maps $Z : \Omega \to H$ given the supremum norm $\|Z\|_\infty := \sup_{\omega\in\Omega} |Z(\omega)|$. Define the map $\mathcal{M} : B(\Omega, H) \to L^0(\Omega, H)$ by

$$
\begin{aligned}
\mathcal{M}(Z)(\omega) := &\int_{-\infty}^{0} T_{-s}p^+ F(Z(\theta(s,\omega)) + Y_1(\theta(s,\omega)))\, ds \\
&- \int_{0}^{\infty} T_{-s}p^- F(Z(\theta(s,\omega)) + Y_1(\theta(s,\omega)))\, ds
\end{aligned}
$$

for all $Z \in B(\Omega, H)$ and all $\omega \in \Omega$.

Note first that $\mathcal{M}$ maps $B(\Omega, H)$ into itself. To see this let $Z \in B(\Omega, H)$ and $\omega \in \Omega$. Then

$$
\begin{aligned}
|\mathcal{M}(Z)(\omega)| &\le \|F\|_\infty \left[\int_{-\infty}^{0} \|T_{-s}p^+\|\, ds + \int_{0}^{\infty} \|T_{-s}p^-\|\, ds\right] \\
&\le \|F\|_\infty \left[\int_{-\infty}^{0} e^{s\mu_{m+1}}\, ds + \int_{0}^{\infty} e^{s\mu_m}\, ds\right] \\
&\le \|F\|_\infty [\mu_{m+1}^{-1} - \mu_m^{-1}] < \infty
\end{aligned}
$$

where $\|F\|_\infty := \sup_{v\in H} |F(v)|$. Hence $\mathcal{M}(Z) \in B(\Omega, H)$ for all $Z \in B(\Omega, H)$.

Secondly, $\mathcal{M}$ is a contraction. To prove this, take any $Z_1, Z_2 \in B(\Omega, H)$ and $\omega \in \Omega$. Then from the definition of $\mathcal{M}$, we get

$$\begin{aligned}
|\mathcal{M}(Z_1)(\omega) - \mathcal{M}(Z_2)(\omega)| &\leq L \int_{-\infty}^{0} \|T_{-s}p^+\| \cdot |Z_1(\theta(s,\omega)) - Z_2(\theta(s,\omega))|\, ds \\
&\quad + L \int_{0}^{\infty} \|T_{-s}p^-\| \cdot |Z_1(\theta(s,\omega)) - Z_2(\theta(s,\omega))|\, ds \\
&\leq L\|Z_1 - Z_2\|_\infty \left[\int_{-\infty}^{0} \|T_{-s}p^+\|\, ds + \int_{0}^{\infty} \|T_{-s}p^-\|\, ds\right] \\
&\leq L\|Z_1 - Z_2\|_\infty \left[\int_{-\infty}^{0} e^{s\mu_{m+1}}\, ds + \int_{0}^{\infty} e^{s\mu_m}\, ds\right] \\
&= L[\mu_{m+1}^{-1} - \mu_m^{-1}]\|Z_1 - Z_2\|_\infty \\
&= \mu\|Z_1 - Z_2\|_\infty
\end{aligned}$$

where $\mu := L[\mu_{m+1}^{-1} - \mu_m^{-1}] < 1$. This proves that $\mathcal{M} : B(\Omega, H) \to B(\Omega, H)$ is a contraction, and hence has a unique fixed point $Z_0 \in B(\Omega, H)$. That is

$$\begin{aligned}
Z_0(\omega) := \int_{-\infty}^{0} & T_{-s}p^+ F(Z_0(\theta(s,\omega)) + Y_1(\theta(s,\omega)))\, ds \\
&- \int_{0}^{\infty} T_{-s}p^- F(Z_0(\theta(s,\omega)) + Y_1(\theta(s,\omega)))\, ds
\end{aligned}$$

for all $\omega \in \Omega$. Now define $Y : \Omega \to H$ by

$$Y(\omega) := Z_0(\omega) + Y_1(\omega), \quad \omega \in \Omega.$$

It is easy to check that Y satisfies the identity (2.4.4).

Since Z_0 is uniquely determined, then so is Y. □

The following proposition gives existence and uniqueness of a stationary point for the see (2.4.1).

PROPOSITION 2.4.2. *Assume all the conditions on A, B_0, F stated in* Proposition 2.4.1. *Suppose that F is globally bounded, globally Lipschitz and satisfies condition* (2.4.3). *Then the semilinear see* (2.4.1) *has a unique stationary point $Y : \Omega \to H$, i.e. $u(t, Y(\omega), \omega) = Y(\theta(t,\omega))$ for all $t \geq 0$ and $\omega \in \Omega$. Furthermore, $Y \in L^p(\Omega, H)$ for all $p \geq 1$.*

PROOF. By hypotheses and Proposition 2.4.1, the integral equation (2.4.4) has a unique $\mathcal{F}$-measurable solution $Y : \Omega \to H$. Let $t \geq 0$, $\omega \in \Omega$. Using (2.4.4), it follows that

$$\begin{aligned}
Y(\theta(t,\omega)) &= \int_{-\infty}^{0} T_{-s}p^+ F(Y(\theta(t+s,\omega)))ds - \int_{0}^{\infty} T_{-s}p^- F(Y(\theta(t+s,\omega)))\, ds \\
&\quad + \left[\int_{-\infty}^{0} T_{-s}p^+ B_0\, dW(s+t) - \int_{0}^{\infty} T_{-s}p^- B_0\, dW(s+t)\right](\omega) \\
&= \int_{-\infty}^{t} T_{t-s}p^+ F(Y(\theta(s,\omega)))ds - \int_{t}^{\infty} T_{t-s}p^- F(Y(\theta(s,\omega)))\, ds \\
&\quad + \left[\int_{-\infty}^{t} T_{t-s}p^+ B_0\, dW(s) - \int_{t}^{\infty} T_{t-s}p^- B_0\, dW(s)\right](\omega)
\end{aligned}$$

$$
\begin{aligned}
=&T_t\Big\{\int_{-\infty}^0 T_{-s}p^+F(Y(\theta(s,\omega)))ds - \int_0^\infty T_{-s}p^-F(Y(\theta(s,\omega)))\,ds\\
&+\Big[\int_{-\infty}^0 T_{-s}p^+B_0\,dW(s) - \int_0^\infty T_{-s}p^-B_0\,dW(s)\Big](\omega)\Big\}\\
&+\int_0^t T_{t-s}p^+F(Y(\theta(s,\omega))ds + \int_0^t T_{t-s}p^-F(Y(\theta(s,\omega)))\,ds\\
&+\Big[\int_0^t T_{t-s}p^+B_0\,dW(s) + \int_0^t T_{t-s}p^-B_0\,dW(s)\Big](\omega)\\
=&T_tY(\omega) + \int_0^t T_{t-s}F(Y(\theta(s,\omega)))ds + \Big[\int_0^t T_{t-s}B_0\,dW(s)\Big](\omega).
\end{aligned}
$$

This gives

$$
Y(\theta(t,\omega)) = T_tY(\omega) + \int_0^t T_{t-s}F(Y(\theta(s,\omega)))ds + \Big[\int_0^t T_{t-s}B_0\,dW(s)\Big](\omega)
$$

for all $t \geq 0$, $\omega \in \Omega$. Therefore, $Y(\theta(t,\omega)), t \geq 0, \omega \in \Omega$, is a stationary solution of (2.4.2) (with $x = Y(\omega)$). Since $u(t, Y(\omega), \omega), t \geq 0, \omega \in \Omega$, is also a solution of (2.4.2), then by uniqueness of the solution to (2.4.2), we must have

$$
u(t, Y(\omega), \omega) = Y(\theta(t,\omega))
$$

for all $t \geq 0$ and all $\omega \in \Omega$. Hence Y is a stationary point for the see (2.4.1).

The stationary point for (2.4.1) is unique (within the class of $\mathcal{F}$-measurable maps $\Omega \to H$). To see this, it is sufficient to observe that the above computation shows that every stationary point of (2.4.1) is a solution of the integral equation (2.4.4). Uniqueness of the stationary solution then follows from Proposition 2.4.1.

In view of the proof of Proposition 2.4.1, the last assertion of Proposition 2.4.2 follows from the fact that $Y_1 \in L^p(\Omega, H)$ for all $p \geq 1$ and $Z_0 \in L^\infty(\Omega, H)$. □.

The existence of local stable and unstable manifolds near a stationary point of the affine stochastic evolution equation (2.4.1) follows from a straightforward modification of the proof of Theorem 2.4.1 in the next section.

(b) Semilinear see's: Linear noise

Here we recall the setting and hypotheses leading to Theorem 1.2.6.

We will prove the existence of local stable and unstable manifolds for semiflows generated by mild solutions of semilinear see's of the form:

$$
\left.\begin{aligned}
du(t) &= -Au(t)dt + F(u(t))dt + Bu(t)\,dW(t), \quad t > 0,\\
u(0) &= x \in H.
\end{aligned}\right\} \tag{2.4.5}
$$

In the above equation $A : D(A) \subset H \to H$ is a closed linear operator on a separable real Hilbert space H. Assume that A has a complete orthonormal system of eigenvectors $\{e_n : n \geq 1\}$ with corresponding positive eigenvalues $\{\mu_n, n \geq 1\}$; i.e., $Ae_n = \mu_n e_n$, $n \geq 1$. Suppose $-A$ generates a strongly continuous semigroup of bounded linear operators $T_t : H \to H$, $t \geq 0$. Let E be a separable real Hilbert space. Suppose $W(t), t \geq 0$, is E-valued cylindrical Brownian motion defined on the canonical complete filtered Wiener space $(\Omega, \bar{\mathcal{F}}, (\mathcal{F}_t)_{t\geq 0}, P)$ and with a separable covariance Hilbert space K, where $K \subset E$ is a Hilbert-Schmidt embedding. That is, Ω is the space of all continuous paths $\omega : \mathbf{R} \to E$ such that $\omega(0) = 0$ with the compact open topology, $\mathcal{F}$ is its Borel σ-field, $\mathcal{F}_t$ is the sub-σ-field generated by all evaluations $\Omega \ni \omega \mapsto \omega(u) \in E, u \leq t$, and P is Wiener measure on Ω. The Brownian motion is given by

$$W(t,\omega) := \omega(t), \quad \omega \in \Omega, t \in \mathbf{R},$$

and may be represented by

$$W(t) = \sum_{k=1}^{\infty} W^k(t) f_k, \quad t \in \mathbf{R},$$

where $\{f_k : k \geq 1\}$ is a complete orthonormal basis of K, and the $W^k, k \geq 1$, are standard independent one-dimensional Wiener processes with $W^k(0) = 0$, $k \geq 1$ ([D-Z.1], Chapter 4).

Suppose $B : H \to L_2(K, H)$ is a bounded linear operator. The stochastic integral in (2.4.5) is defined in the sense of ([D-Z.1], Chapter 4).

Assume the hypotheses of Theorem 1.2.4.

We will denote by $\theta : \mathbf{R} \times \Omega \to \Omega$ the standard P-preserving ergodic Wiener shift on Ω:

$$\theta(t,\omega)(s) := \omega(t+s) - \omega(t), \quad t, s \in \mathbf{R}.$$

Let $L(H)$ be the Banach space of all bounded linear operators $H \to H$ given the uniform operator norm $\|\cdot\|$. Denote by $L_2(H) \subset L(H)$ the Hilbert space of all Hilbert-Schmidt operators $S : H \to H$.

Suppose $F : H \to H$ is a (Fréchet) $C^{k,\epsilon}$ $(k \geq 1, \epsilon \in (0,1])$ non-linear map satisfying the following Lipschitz and linear growth hypotheses:

$$\left.\begin{aligned} |F(v)| &\leq C(1+|v|), \quad v \in H \\ |F(v_1) - F(v_2)| &\leq L_n|v_1 - v_2|, \quad v_i \in H, |v_i| \leq n, i = 1, 2, \end{aligned}\right\} \tag{2.4.6}$$

for some positive constants $C, L_n, n \geq 1$.

The mild solutions of the see (2.4.5) generate a $C^{k,\epsilon}$ $(k \geq 1, \epsilon \in (0,1])$ perfect cocycle (U, θ) on H, satisfying all the assertions of Theorem 1.2.6.

Under the above conditions, one gets the following stable manifold theorem for hyperbolic stationary trajectories of the see (2.4.5).

THEOREM 2.4.1. *Assume the above hypotheses on the coefficients of the see* (2.4.5). *Assume that the stochastic semiflow* $U : \mathbf{R}^+ \times H \times \Omega \to H$ *generated by mild solutions of* (2.4.5) *has a hyperbolic stationary point* $Y : \Omega \to H$ *such that* $E \log^+ |Y| < \infty$. *Then* (U, θ) *has a perfect family of* $C^{k,\epsilon}$ *local stable and unstable manifolds satisfying all the assertions of* Theorem 2.2.1.

PROOF. One first checks the estimate

$$\int_\Omega \log^+ \sup_{0 \leq t_1, t_2 \leq a} \|U(t_2, Y(\theta(t_1,\omega)) + (\cdot), \theta(t_1,\omega))\|_{k,\epsilon}\, dP(\omega) < \infty \tag{2.4.7}$$

for any fixed $0 < \rho, a < \infty, k \geq 1$ and $\epsilon \in (0,1]$. This estimate follows from the integrability condition on Y and assertion (vi) of Theorem 1.2.6. The conclusion of Theorem 2.4.1 now follows immediately from Theorem 2.2.1. □

(c) Semilinear parabolic spde's: Lipschitz nonlinearity

Consider the Laplacian

$$\Delta := \frac{1}{2}\sum_{i,j=1}^{d} \frac{\partial^2}{\partial \xi_i^2} \tag{2.4.8}$$

defined on a smooth bounded domain $\mathcal{D}$ in $\mathbf{R}^d$, with a smooth boundary $\partial\mathcal{D}$ with zero Dirichlet boundary conditions. Assume that $f : \mathbf{R} \to \mathbf{R}$ is a C_b^∞ function and let $d\xi$ be Lebesgue measure on $\mathbf{R}^d$. Let $W^i, i \geq 1$, be independent one-dimensional standard Brownian motions with $W^i(0) = 0$ defined on the canonical complete filtered Wiener space $(\Omega, \bar{\mathcal{F}}, (\mathcal{F}_t)_{t\in\mathbf{R}}, P)$. Let θ denote the Brownian shift on $\Omega := C(\mathbf{R}, \mathbf{R}^\infty)$. Recall that the Sobolev space $H_0^k(\mathcal{D})$ is the completion of $C_0^\infty(\mathcal{D}, \mathbf{R})$ under the Sobolev norm

$$||u||^2_{H_0^k(\mathcal{D})} := \sum_{|\alpha|\leq k} \int_{\mathcal{D}} |D^\alpha u(\xi)|^2 \, d\xi.$$

Suppose further that $\sigma_i \in H_0^s(\mathcal{D}), i \geq 1$, and the series $\sum_{i=1}^{\infty} ||\sigma_i||^2_{H_0^s}$ converges, where $s > k + \frac{d}{2} > d$.

By Theorem 1.3.5, weak solutions of the initial-value problem:

$$\left.\begin{aligned} du(t) &= \frac{1}{2}\Delta u(t)dt + f(u(t))dt + \sum_{i=1}^{\infty} \sigma_i u(t)\, dW^i(t), \quad t > 0 \\ u(0) &= \psi \in H_0^k(\mathcal{D}) \end{aligned}\right\} \tag{2.4.9}$$

give a perfect smooth cocycle (U, θ) on the Sobolev space $H_0^k(\mathcal{D})$ which satisfies all the assertions of Theorem 1.3.5. Applying Theorem 2.2.1, we get the following stable manifold theorem for the spde (2.4.9):

THEOREM 2.4.2. *Assume the above hypotheses on the coefficients of the spde* (2.4.9). *Assume that the stochastic semiflow* $U : \mathbf{R}^+ \times H_0^k(\mathcal{D}) \times \Omega \to H_0^k(\mathcal{D})$ *generated by weak solutions of* (2.4.9) *has a hyperbolic stationary point* $Y : \Omega \to H_0^k(\mathcal{D})$ *such that* $E \log^+ ||Y||_{H_0^k} < \infty$. *Then* (U, θ) *has a perfect family of* C^∞ *local stable and unstable manifolds in* $H_0^k(\mathcal{D})$ *satisfying all the assertions of* Theorem 2.2.1.

(d) Stochastic reaction diffusion equations: Dissipative nonlinearity

In Section 1.4 (a), we constructed a C^1 stochastic semiflow on the Hilbert space $H := L^2(\mathcal{D})$ for a stochastic reaction-diffusion equation

$$du = \nu\Delta u \, dt + u(1 - |u|^\alpha)\, dt + \sum_{i=1}^{\infty} \sigma_i u(t)\, dW^i(t), \tag{2.4.10}$$

defined on a bounded domain $\mathcal{D} \subset \mathbf{R}^d$ with a smooth boundary $\partial\mathcal{D}$. In (2.4.10), the Laplacian on $\mathcal{D}$ is denoted by Δ, and we impose Dirichlet boundary conditions on $\partial\mathcal{D}$. The $W^i, i \geq 1$, are independent one-dimensional standard Brownian motions as in Section 1.4 (a), and the series $\sum_{i=1}^{\infty} \|\sigma_i\|^2_{H_0^s}$ converges for $s > 2+\frac{d}{2}$. The dissipative term yields the existence of a unique stationary solution of (2.4.10) under a suitable choice of the parameter ν ([D-Z.2]).

In view of the estimates in Theorem 1.4.1 and Theorem 2.2.1, one gets the following:

THEOREM 2.4.3. *Assume the above hypotheses on the coefficients of the spde* (2.4.10). *Let* $\alpha < \frac{4}{d}$. *Assume that the stochastic semiflow* $U : \mathbf{R}^+ \times L^2(\mathcal{D}) \times \Omega \to L^2(\mathcal{D})$ *generated by mild solutions of* (2.4.10) *has a hyperbolic stationary point* $Y : \Omega \to L^2(\mathcal{D})$ *such that* $E \log^+ \|Y\|_{L^2} < \infty$. *Then* (U, θ) *has a perfect family of* C^1 *local stable and unstable manifolds in* $L^2(\mathcal{D})$ *satisfying all the assertions of Theorem 2.2.1.*

REMARKS.

(i) The results in Sections (c) and (d) hold if the Euclidean domain $\mathcal{D}$ is replaced by a compact smooth d-dimensional Riemannian manifold M (possibly with a smooth boundary ∂M).

(ii) We conjecture that Theorem 2.4.3 still holds (but with *Lipschitz* stable/unstable manifolds) if the dissipative term $u(1-|u|^\alpha)$ is replaced by a more general one of the form $F(u) := f \circ u$, where $f : \mathbf{R} \to \mathbf{R}$ is a C^1 function satisfying the following classical estimates:

$$-c_1 - \alpha_1|x|^p \le f(x)x \le c_1 - \alpha_2|x|^p, \quad f'(x) \le c_2,$$

for all $x \in \mathbf{R}$, with $c_1, c_2, \alpha_1, \alpha_2$ positive constants, and p any integer greater than 2.

(iii) Is it true that the stochastic flow and the local stable/unstable manifolds in Theorem 2.4.3 are of class C^2?

(e) Stochastic Burgers equation: Additive noise

The existence of a C^1 stochastic semiflow on $L^2([0,1])$ for Burgers equation

$$du + u\frac{\partial u}{\partial \xi}\,dt = \nu\Delta u dt + dW(t), \quad t > 0,\ \nu > 0, \tag{2.4.11}$$

was established in Part 1 of this paper, where $W(t)$, $t > 0$, is an infinite dimensional Brownian motion on $L^2[0,1]$. See Theorem 1.4.3.

Under extra spatial smoothness hypotheses on the noise, namely $W(t,\cdot) \in C^3([0,1])$, Burgers equation (2.4.11) admits a unique stationary point ([Si]). More generally, with our weaker condition on the noise W (Section 1.4 (b)), we stipulate that equation (2.4.11) has a hyperbolic stationary point. In this case, we get the following result:

THEOREM 2.4.4. *Assume the hypotheses of* Theorem 1.4.3 *on the coefficients of Burgers spde* (2.4.11). *Assume that the stochastic semiflow* $U : \mathbf{R}^+ \times L^2([0,1]) \times \Omega \to L^2([0,1])$ *generated by mild solutions of* (2.4.11) *has a hyperbolic stationary point* $Y : \Omega \to L^2([0,1])$ *such that* $E \log^+ \|Y\|_{L^2} < \infty$. *Then* (U, θ) *has a perfect family of* C^1 *local stable and unstable manifolds in* $L^2([0,1])$ *satisfying all the assertions of* Theorem 2.2.1.

Note that hyperbolicity of the stationary point in Theorem 2.4.4 is in the sense of Definition (2.1.3). Theorems 2.1.1 and 1.4.3 imply that the Lyapunov spectrum for the linearization of (2.4.11) exists and is discrete for any viscosity $\nu > 0$. When W is C^3 in the space variable, it known that for any C^2 initial condition, the solution $u(t)$ of (2.4.11) converges to the stationary solution for any positive viscosity $\nu > 0$ ([Si]). It is therefore easy to see that the stable manifold is the whole of $L^2([0,1])$.

The case of sufficiently large viscosity and rough noise $W(t) \in L^2([0,1])$ is currently being studied ([L-Z]). This work shows that (2.4.11) admits a unique globally exponentially stable stationary point in this case. So in this (somewhat non-generic) case, the unstable manifold consists of the single random point $Y(\omega) \in L^2([0,1])$, and the non-linear cocycle U will approach $Y(\omega)$ with exponential speed less than or equal to the top Lyapunov exponent λ_1 of the linearized Burgers equation.

We conjecture that the assertions in the above paragraph still hold for any viscosity $\nu > 0$ (cf. [D-Z.2], Theorem 14.4.4). Further analysis of the Lyapunov spectrum for (2.4.11) (in the cases of small and zero viscosity ν) is postponed to a future project.

Acknowledgments

The authors would like to thank Prof. B. Øksendal for inviting them to Oslo in the summer of 2000 where the project was started. They are also grateful to Prof. K.D. Elworthy for inviting them to Warwick on numerous occasions especially during the Warwick SPDE's Symposium 2000/2001 so that they may have opportunities to meet; to Prof. A. Truman for inviting them to the International Workshop of Probabilistic Methods in Fluids at Swansea in April 2002 where preliminary versions of the results were announced; and to the organizers of ICM 2002 and First Sino-German Stochastic Analysis Conference in Beijing in August 2002 where the results were presented. The authors would also like to thank the referees for useful comments.

Bibliography

A. L. Arnold, *Random dynamical systems*, Springer-Verlag, Berlin, 1998.

A-S. L. Arnold and M. K. R. Scheutzow, *Perfect cocycles through stochastic differential equations*, Probab. Th. Rel. Fields **101** (1995), 65–88.

B-C-F. Z. Brzezniak, M. Capinski, and F. Flandoli, *Stochastic Navier-Stokes equations with multiplicative noise*, Stochastic Analysis and Applications **10** (1992), 53–532.

B-C-J. L. Bertini, N. Cancrin, and G. Jona-Lasinio, *The stochastic Burgers equation*, Comm. Math. Phys. **165** (1994), 211–232.

B-F. Z. Brzezniak and F. Flandoli, *Regularity of solutions and random evolution operator for stochastic parabolic equations*, Stochastic partial differential equations and applications (Trento, 1990), Pitman Res. Notes Math. Ser., vol. 268, Longman Sci. Tech., Harlow, 1992, pp. 54–71.

B-F.1. A. Bensoussan and F. Flandoli, *Stochastic inertial manifold*, Stochastics Stochastics Rep. **53** (1995), no. 1–2, 13–39.

C-K-S. T. Caraballo, P. E. Kloeden, and B. Schmalfuss, *Exponentially stable stationary solution for stochastic evolution equations and their perturbation*, Appl. Math. Optim. **50** (2004), 183–207.

C. A. Carverhill, *Flows of stochastic dynamical systems: Ergodic theory*, Stochastics **14** (1985), 273–317.

C-V. C. Castaing and M. Valadier, *Convex Analysis and Measurable Multifunctions*, Lecture Notes in Mathematics, vol. 580, Springer-Verlag, Berlin-Heidelberg-New York, 1977.

Co. D. L. Cohn, *Measure Theory*, Birkhäuser, Boston, 1980.

D-D-T. G. Da Prato, A. Debussche, and R. Temam, *Stochastic Burgers equation*, Nonlinear Differential Eq. Appl. **1** (1994), 389.

D-Z.1. G. Da Prato and J. Zabczyk, *Stochastic Equations in Infinite Dimensions*, Cambridge University Press, Cambridge, 1992.

D-Z.2. G. Da Prato and J. Zabczyk, *Ergodicity for Infinite Dimensional Systems*, Cambridge University Press, Cambridge, 1996.

D-T-Z. I. M. Davies, A. Truman, and H. Z. Zhao, *Stochastic generalized KPP equations*, Proc. R. Soc. Edinb. **126A** (1996), no. 5, 957–84.

D-L-S.1. J. Duan, K. Lu, and B. Schmalfuss, *Invariant manifolds for stochastic partial differential equations*, Annals of Probability **31** (2003), 2109–2135.

D-L-S.2. J. Duan, K. Lu, and B. Schmalfuss, *Stable and unstable manifolds for stochastic partial differential equations*, J. Dynamics and Diff. Eqns. **16** (2004), no. 4, 949–972.

E-V. E. W. Vanden and E. Eijnden, *Statistical theory for the stochastic Burgers equation in the inviscid limit*, Comm. Pure Appl. Math. **53** (2000), 852–901.

E-H. J. -P. Eckmann and M. Hairer, *Invariant measures for stochastic partial differential equations in unbounded domains*, Nonlinearity **14** (2001), 133–151.

E-Z. K. D. Elworthy and H. Z. Zhao, *The propagation of travelling waves for stochastic generalized KPP equations*, Mathl. and Comput. Modelling **20** (1994), no. 4/5, 131–166.

F.1. F. Flandoli, *Stochastic flows for nonlinear second-order parabolic SPDE's*, Stochastics Monographs, 9, Gordon and Breach Science Publishers, Yverdon, 1995.

F.2. F. Flandoli, *Stochastic flows for nonlinear second-order parabolic SPDEs*, Ann. Probab. **24** (1996), no. 2, 547–558.

F-S.1. F. Flandoli and K.-U. Schaumlöffel, *Stochastic parabolic equations in bounded domains: random evolution operator and Lyapunov exponents*, Stochastics and Stochastics Reports **29** (1990), no. 4, 461–485.

F-S.2. F. Flandoli and K.-U. Schaumlöffel, *A multiplicative ergodic theorem with applications to a first order stochastic hyperbolic equation in a bounded domain*, Stochastics Stochastics Rep. **34** (1991), no. 3–4, 241–255.

Fr. M. Freidlin, *Functional integration and partial differential equations*, Princeton University Press, Princeton, NJ, 1985.

Fri. A. Friedman, *Partial differential equations of parabolic type*, Prentice-Hall, Inc., Englewood Cliffs, NJ, 1964.

H-L-O-U-Z. H. Holden, T. Lindstrom, B. Øksendal, J. Uboe, and T. S. Zhang, *The Burgers' cquation with a noise force and the stochastic heat equations*, Comm. PDE **19** (1994), 119–141.

I-W. N. Ikeda and S. Watanabe, *Stochastic Differential Equations and Diffusion Processes*, Second Edition, North-Holland-Kodansha, Amsterdam, 1989.

Ku. H. Kunita, *Stochastic Flows and Stochastic Differential Equations*, Cambridge University Press, Cambridge, New York, Melbourne, Sydney, 1990.

L-S-U. O. A. Ladyzenskaja, V. A. Solonnikov, and N. N. Uralceva, *Linear and quasi-linear equations of parabolic type*, Translations of Mathematical Monographs, vol. 23, American Mathematical Society, Providence, RI, 1968.

L-Z. Y. Liu, and H. Z. Zhao, *The stochastic Burgers integral equations and the stationary solutions of stochastic Burgers equations*, in preparation.

Ly. M. A. Lyapunov, *Problèm général de la stablité du mouvement*, Annales Dac. Sciences Toulouse **9** (1907); (Translation of the Russian edition, Kharkov 1892). Reprinted by Princeton University Press, Princeton, N.J. 1949 and 1952.

Mo.1. S.-E. A. Mohammed, *Stochastic Functional Differential Equations*, Research Notes in Mathematics, no. 99, Pitman Advanced Publishing Program, Boston-London-Melbourne, 1984.

Mo.2. S.-E. A. Mohammed, *Non-Linear flows for linear stochastic delay equations*, Stochastics, **17** (1987), no. 3, 207–212.

Mo.3. S.-E. A. Mohammed, *The Lyapunov spectrum and stable manifolds for stochastic linear delay equations*, Stochastics and Stochastic Reports **29** (1990), 89–131.

M-S.1. S.-E. A. Mohammed and M. K. R. Scheutzow, *The stable manifold theorem for non-linear stochastic systems with memory. Part I: Existence of the semiflow. Part II: The local stable manifold theorem*, preprints (2001).

M-S.2. S.-E. A. Mohammed and M. K. R. Scheutzow, *The stable manifold theorem for stochastic differential equations*, The Annals of Probability **27** (1999), no. 2, 615–652.

M-S.3. S.-E. A. Mohammed and M. K. R. Scheutzow, *Spatial estimates for stochastic flows in Euclidean space*, The Annals of Probability **26** (1998), no. 1, 56–77.

M-S.4. S.-E. A. Mohammed and M. K. R. Scheutzow, *Lyapunov exponents of linear and stochastic functional differential equations driven by semimartingales, Part I: The multiplicative ergodic theory*, Annals of Institute of Henri Poincare **32** (1996), no. 1, 69–105.

O-V-Z. B. Øksendal, G. Väge, and H. Z. Zhao, *Two properties of stochastic KPP equations: ergodicity and pathwise property*, Nonlinearity **14** (2001), 639–662.

O. Oseledec, V. I, *A multiplicative ergodic theorem. Lyapunov characteristic numbers for dynamical systems*, Trudy Moskov. Mat. Obšč. **19** (1968), 179–210; English translation, Trans. Moscow Math. Soc. **19** (1968), 197–221.

Pr. Ph. E. Protter, *Stochastic Integration and Stochastic Differential Equations: A New Approach*, Springer, Berlin, 1990.

Ro. J. C. Robinson, *Infinite-Dimensional Dynamical Systems. An Introduction to Dissipative Parabolic PDE's and the Theory of Global Attractors*, Cambridge Texts in Applied Mathematics, Cambridge University Press, Cambridge, 2001.

Ru.1. D. Ruelle, *Ergodic theory of differentiable dynamical systems*, Publ. Math. Inst. Hautes Etud. Sci. (1979), 275–306.

Ru.2. D. Ruelle, *Characteristic exponents and invariant manifolds in Hilbert space*, Annals of Math. **115** (1982), 243–290.

Sc. B. Schmalfuss, *A random fixed point theorem and the random graph transformation*, J. Math. Anal. Appli. **225** (1998), 91–113.

Si. Ya. G. Sinai, *Burgers system driven by a periodic stochastic flow*, Itô's stochastic calculus and probability theory, Springer, Tokyo, 1996, pp. 347–353.

Sk. A. V. Skorohod, *Random Linear Operators*, Riedel, 1984.

Ta. M. E. Taylor, *Partial Differential Equations III Nonlinear Equations*, Springer, New York, 1996.

Te. R. Temam, *Infinite-Dimensional Dynamical Systems in Mechanics and Physics*, Springer-Verlag, New York, 1988.

Tw. K. Twardowska, *An extension of the Wong-Zakai theorem for stochastic equations in Hilbert spaces*, Stochastic Anal. Appl. **10** (1992), no. 4, 471–500.

T-Za. R. Tribe and O. Zaboronski, *On the large time asymptotics of decaying Burgers turbulence*, Comm. Math. Phys. **212** (2000), 415–436.

T-Z. A. Truman and H. Z. Zhao, *Stochastic Burgers' equations and their semi-classical expansions*, Comm. Math. Phys. **194** (1998), 231–248.

Wa. J. B. Walsh, *An Introduction to stochastic partial differential equations*, École d'eté de Probabilité de Saint Flour, XIV, 1984,, ed. P.L. Hennequim, Lecture Notes in Mathematics, no. 1180, Springer-Verlag, Berlin, 1986, pp. 265-439.

Wan. T. Wanner, *Linearization of random dynamical systems*, Dynamics Reported, vol. 4, Springer-Verlag, Berlin-Heidelberg-New York, 1995, pp. 203-269.

Z-Z. Q. Zhang and H. Z. Zhao, *Pathwise stationary solutions of stochastic partial differential equations and backward doubly stochastic differential equations on infinite horizon*, preprint.

Editorial Information

To be published in the *Memoirs*, a paper must be correct, new, nontrivial, and significant. Further, it must be well written and of interest to a substantial number of mathematicians. Piecemeal results, such as an inconclusive step toward an unproved major theorem or a minor variation on a known result, are in general not acceptable for publication.

Papers appearing in *Memoirs* are generally at least 80 and not more than 200 published pages in length. Papers less than 80 or more than 200 published pages require the approval of the Managing Editor of the Transactions/Memoirs Editorial Board.

As of July 31, 2008, the backlog for this journal was approximately 16 volumes. This estimate is the result of dividing the number of manuscripts for this journal in the Providence office that have not yet gone to the printer on the above date by the average number of monographs per volume over the previous twelve months, reduced by the number of volumes published in four months (the time necessary for preparing a volume for the printer). (There are 6 volumes per year, each usually containing at least 4 numbers.)

A Consent to Publish and Copyright Agreement is required before a paper will be published in the *Memoirs*. After a paper is accepted for publication, the Providence office will send a Consent to Publish and Copyright Agreement to all authors of the paper. By submitting a paper to the *Memoirs*, authors certify that the results have not been submitted to nor are they under consideration for publication by another journal, conference proceedings, or similar publication.

Information for Authors

Memoirs are printed from camera copy fully prepared by the author. This means that the finished book will look exactly like the copy submitted.

Initial submission. The AMS uses Centralized Manuscript Processing for initial submissions. Authors should submit a PDF file using the Initial Manuscript Submission form found at `www.ams.org/peer-review-submission`, or send one copy of the manuscript to the following address: Centralized Manuscript Processing, MEMOIRS OF THE AMS, 201 Charles Street, Providence, RI 02904-2294 USA. If a paper copy is being forwarded to the AMS, indicate that it is for it Memoirs and include the name of the corresponding author, contact information such as email address or mailing address, and the name of an appropriate Editor to review the paper (see the list of Editors below).

The paper must contain a *descriptive title* and an *abstract* that summarizes the article in language suitable for workers in the general field (algebra, analysis, etc.). The *descriptive title* should be short, but informative; useless or vague phrases such as "some remarks about" or "concerning" should be avoided. The *abstract* should be at least one complete sentence, and at most 300 words. Included with the footnotes to the paper should be the 2000 *Mathematics Subject Classification* representing the primary and secondary subjects of the article. The classifications are accessible from `www.ams.org/msc/`. The list of classifications is also available in print starting with the 1999 annual index of *Mathematical Reviews*. The Mathematics Subject Classification footnote may be followed by a list of *key words and phrases* describing the subject matter of the article and taken from it. Journal abbreviations used in bibliographies are listed in the latest *Mathematical Reviews* annual index. The series abbreviations are also accessible from `www.ams.org/msnhtml/serials.pdf`. To help in preparing and verifying references, the AMS offers MR Lookup, a Reference Tool for Linking, at `www.ams.org/mrlookup/`.

Electronically prepared manuscripts. The AMS encourages electronically prepared manuscripts, with a strong preference for $\mathcal{A}\mathcal{M}\mathcal{S}$-LaTeX. To this end, the Society has prepared $\mathcal{A}\mathcal{M}\mathcal{S}$-LaTeX author packages for each AMS publication. Author packages include instructions for preparing electronic manuscripts, samples, and a style file that generates the particular design specifications of that publication series. Though $\mathcal{A}_\mathcal{M}\mathcal{S}$-LaTeX is the highly preferred format of TeX, author packages are also available in $\mathcal{A}_\mathcal{M}\mathcal{S}$-TeX.

Authors may retrieve an author package for *Memoirs of the AMS* from `www.ams.org/journals/memo/memoauthorpac.html` or via FTP to `ftp.ams.org` (login as `anonymous`, enter username as password, and type `cd pub/author-info`). The *AMS Author Handbook* and the *Instruction Manual* are available in PDF format from the author package link. The author package can also be obtained free of charge by sending email to `tech-support@ams.org` (Internet) or from the Publication Division, American Mathematical Society, 201 Charles St., Providence, RI 02904-2294, USA. When requesting an author package, please specify $\mathcal{A}_\mathcal{M}\mathcal{S}$-LaTeX or $\mathcal{A}_\mathcal{M}\mathcal{S}$-TeX and the publication in which your paper will appear. Please be sure to include your complete mailing address.

After acceptance. The final version of the electronic file should be sent to the Providence office (this includes any TeX source file, any graphics files, and the DVI or PostScript file) immediately after the paper has been accepted for publication.

Before sending the source file, be sure you have proofread your paper carefully. The files you send must be the EXACT files used to generate the proof copy that was accepted for publication. For all publications, authors are required to send a printed copy of their paper, which exactly matches the copy approved for publication, along with any graphics that will appear in the paper.

Accepted electronically prepared files can be submitted via the web at `www.ams.org/submit-book-journal/`, sent via FTP, or sent on CD-Rom or diskette to the Electronic Prepress Department, American Mathematical Society, 201 Charles Street, Providence, RI 02904-2294 USA. TeX source files, DVI files, and PostScript files can be transferred over the Internet by FTP to the Internet node `ftp.ams.org` (130.44.1.100). When sending a manuscript electronically via CD-Rom or diskette, please be sure to include a message identifying the paper as a Memoir.

Electronically prepared manuscripts can also be sent via email to `pub-submit@ams.org` (Internet). In order to send files via email, they must be encoded properly. (DVI files are binary and PostScript files tend to be very large.)

Electronic graphics. Comprehensive instructions on preparing graphics are available at `www.ams.org/authors/journals.html`. A few of the major requirements are given here.

Submit files for graphics as EPS (Encapsulated PostScript) files. This includes graphics originated via a graphics application as well as scanned photographs or other computer-generated images. If this is not possible, TIFF files are acceptable as long as they can be opened in Adobe Photoshop or Illustrator. No matter what method was used to produce the graphic, it is necessary to provide a paper copy to the AMS.

Authors using graphics packages for the creation of electronic art should also avoid the use of any lines thinner than 0.5 points in width. Many graphics packages allow the user to specify a "hairline" for a very thin line. Hairlines often look acceptable when proofed on a typical laser printer. However, when produced on a high-resolution laser imagesetter, hairlines become nearly invisible and will be lost entirely in the final printing process.

Screens should be set to values between 15% and 85%. Screens which fall outside of this range are too light or too dark to print correctly. Variations of screens within a graphic should be no less than 10%.

Inquiries. Any inquiries concerning a paper that has been accepted for publication should be sent to `memo-query@ams.org` or directly to the Electronic Prepress Department, American Mathematical Society, 201 Charles St., Providence, RI 02904-2294 USA.

Editors

This journal is designed particularly for long research papers, normally at least 80 pages in length, and groups of cognate papers in pure and applied mathematics. Papers intended for publication in the *Memoirs* should be addressed to one of the following editors. The AMS uses Centralized Manuscript Processing for initial submissions to AMS journals. Authors should follow instructions listed on the Initial Submission page found at `www.ams.org/memo/memosubmit.html`.

Algebra to ALEXANDER KLESHCHEV, Department of Mathematics, University of Oregon, Eugene, OR 97403-1222; email: `ams@noether.uoregon.edu`

Algebraic geometry and its application to MINA TEICHER, Emmy Noether Research Institute for Mathematics, Bar-Ilan University, Ramat-Gan 52900, Israel; email: `teicher@macs.biu.ac.il`

Algebraic geometry to DAN ABRAMOVICH, Department of Mathematics, Brown University, Box 1917, Providence, RI 02912; email: `amsedit@math.brown.edu`

Algebraic topology to ALEJANDRO ADEM, Department of Mathematics, University of British Columbia, Room 121, 1984 Mathematics Road, Vancouver, British Columbia, Canada V6T 1Z2; email: `adem@math.ubc.ca`

Combinatorics to JOHN R. STEMBRIDGE, Department of Mathematics, University of Michigan, Ann Arbor, Michigan 48109-1109; email: `FRS@umich.edu`

Complex analysis and harmonic analysis to ALEXANDER NAGEL, Department of Mathematics, University of Wisconsin, 480 Lincoln Drive, Madison, WI 53706-1313; email: `nagel@math.wisc.edu`

Differential geometry and global analysis to LISA C. JEFFREY, Department of Mathematics, University of Toronto, 100 St. George St., Toronto, ON Canada M5S 3G3; email: `jeffrey@math.toronto.edu`

Dynamical systems and ergodic theory and complex anaysis to YUNPING JIANG, Department of Mathematics, CUNY Queens College and Graduate Center, 65-30 Kissena Blvd., Flushing, NY 11367; email: `Yunping.Jiang@qc.cuny.edu`

Functional analysis and operator algebras to DIMITRI SHLYAKHTENKO, Department of Mathematics, University of California, Los Angeles, CA 90095; email: `shlyakht@math.ucla.edu`

Geometric analysis to WILLIAM P. MINICOZZI II, Department of Mathematics, Johns Hopkins University, 3400 N. Charles St., Baltimore, MD 21218; email: `trans@math.jhu.edu`

Geometric analysis to MARK FEIGHN, Math Department, Rutgers University, Newark, NJ 07102; email: `feighn@andromeda.rutgers.edu`

Harmonic analysis, representation theory, and Lie theory to ROBERT J. STANTON, Department of Mathematics, The Ohio State University, 231 West 18th Avenue, Columbus, OH 43210-1174; email: `stanton@math.ohio-state.edu`

Logic to STEFFEN LEMPP, Department of Mathematics, University of Wisconsin, 480 Lincoln Drive, Madison, Wisconsin 53706-1388; email: `lempp@math.wisc.edu`

Number theory to JONATHAN ROGAWSKI, Department of Mathematics, University of California, Los Angeles, CA 90095; email: `jonr@math.ucla.edu`

Partial differential equations to GUSTAVO PONCE, Department of Mathematics, South Hall, Room 6607, University of California, Santa Barbara, CA 93106; email: `ponce@math.ucsb.edu`

Partial differential equations and dynamical systems to PETER POLACIK, School of Mathematics, University of Minnesota, Minneapolis, MN 55455; email: `polacik@math.umn.edu`

Probability and statistics to RICHARD BASS, Department of Mathematics, University of Connecticut, Storrs, CT 06269-3009; email: `bass@math.uconn.edu`

Real analysis and partial differential equations to DANIEL TATARU, Department of Mathematics, University of California, Berkeley, Berkeley, CA 94720; email: `tataru@math.berkeley.edu`

All other communications to the editors should be addressed to the Managing Editor, ROBERT GURALNICK, Department of Mathematics, University of Southern California, Los Angeles, CA 90089-1113; email: `guralnic@math.usc.edu`.

Titles in This Series

918 **Jonathan Brundan and Alexander Kleshchev,** Representations of shifted Yangians and finite W-algebras, 2008

917 **Salah-Eldin A. Mohammed, Tusheng Zhang, and Huaizhong Zhao,** The stable manifold theorem for semilinear stochastic evolution equations and stochastic partial differential equations, 2008

916 **Yoshikata Kida,** The mapping class group from the viewpoint of measure equivalence theory, 2008

915 **Sergiu Aizicovici, Nikolaos S. Papageorgiou, and Vasile Staicu,** Degree theory for operators of monotone type and nonlinear elliptic equations with inequality constraints, 2008

914 **E. Shargorodsky and J. F. Toland,** Bernoulli free-boundary problems, 2008

913 **Ethan Akin, Joseph Auslander, and Eli Glasner,** The topological dynamics of Ellis actions, 2008

912 **Igor Chueshov and Irena Lasiecka,** Long-time behavior of second order evolution equations with nonlinear damping, 2008

911 **John Locker,** Eigenvalues and completeness for regular and simply irregular two-point differential operators, 2008

910 **Joel Friedman,** A proof of Alon's second eigenvalue conjecture and related problems, 2008

909 **Cameron McA. Gordon and Ying-Qing Wu,** Toroidal Dehn fillings on hyperbolic 3-manifolds, 2008

908 **J.-L. Waldspurger,** L'endoscopie tordue n'est pas si tordue, 2008

907 **Yuanhua Wang and Fei Xu,** Spinor genera in characteristic 2, 2008

906 **Raphaël S. Ponge,** Heisenberg calculus and spectral theory of hypoelliptic operators on Heisenberg manifolds, 2008

905 **Dominic Verity,** Complicial sets characterising the simplicial nerves of strict ω-categories, 2008

904 **William M. Goldman and Eugene Z. Xia,** Rank one Higgs bundles and representations of fundamental groups of Riemann surfaces, 2008

903 **Gail Letzter,** Invariant differential operators for quantum symmetric spaces, 2008

902 **Bertrand Toën and Gabriele Vezzosi,** Homotopical algebraic geometry II: Geometric stacks and applications, 2008

901 **Ron Donagi and Tony Pantev (with an appendix by Dmitry Arinkin),** Torus fibrations, gerbes, and duality, 2008

900 **Wolfgang Bertram,** Differential geometry, Lie groups and symmetric spaces over general base fields and rings, 2008

899 **Piotr Hajłasz, Tadeusz Iwaniec, Jan Malý, and Jani Onninen,** Weakly differentiable mappings between manifolds, 2008

898 **John Rognes,** Galois extensions of structured ring spectra/Stably dualizable groups, 2008

897 **Michael I. Ganzburg,** Limit theorems of polynomial approximation with exponential weights, 2008

896 **Michael Kapovich, Bernhard Leeb, and John J. Millson,** The generalized triangle inequalities in symmetric spaces and buildings with applications to algebra, 2008

895 **Steffen Roch,** Finite sections of band-dominated operators, 2008

894 **Martin Dindoš,** Hardy spaces and potential theory on C^1 domains in Riemannian manifolds, 2008

893 **Tadeusz Iwaniec and Gaven Martin,** The Beltrami Equation, 2008

892 **Jim Agler, John Harland, and Benjamin J. Raphael,** Classical function theory, operator dilation theory, and machine computation on multiply-connected domains, 2008

891 **John H. Hubbard and Peter Papadopol,** Newton's method applied to two quadratic equations in $\mathbb{C}^2$ viewed as a global dynamical system, 2008

890 **Steven Dale Cutkosky,** Toroidalization of dominant morphisms of 3-folds, 2007

889 **Michael Sever,** Distribution solutions of nonlinear systems of conservation laws, 2007

888 **Roger Chalkley,** Basic global relative invariants for nonlinear differential equations, 2007

887 **Charlotte Wahl,** Noncommutative Maslov index and eta-forms, 2007

886 **Robert M. Guralnick and John Shareshian,** Symmetric and alternating groups as monodromy groups of Riemann surfaces I: Generic covers and covers with many branch points, 2007

885 **Jae Choon Cha,** The structure of the rational concordance group of knots, 2007

884 **Dan Haran, Moshe Jarden, and Florian Pop,** Projective group structures as absolute Galois structures with block approximation, 2007

883 **Apostolos Beligiannis and Idun Reiten,** Homological and homotopical aspects of torsion theories, 2007

882 **Lars Inge Hedberg and Yuri Netrusov,** An axiomatic approach to function spaces, spec tral synthesis and Luzin approximation, 2007

881 **Tao Mei,** Operator valued Hardy spaces, 2007

880 **Bruce C. Berndt, Geumlan Choi, Youn-Seo Choi, Heekyoung Hahn, Boon Pin Yeap, Ae Ja Yee, Hamza Yesilyurt, and Jinhee Yi,** Ramanujan's forty identities for Rogers-Ramanujan functions, 2007

879 **O. García-Prada, P. B. Gothen, and V. Muñoz,** Betti numbers of the moduli space of rank 3 parabolic Higgs bundles, 2007

878 **Alessandra Celletti and Luigi Chierchia,** KAM stability and celestial mechanics, 2007

877 **María J. Carro, José A. Raposo, and Javier Soria,** Recent developments in the theory of Lorentz spaces and weighted inequalities, 2007

876 **Gabriel Debs and Jean Saint Raymond,** Borel liftings of Borel sets: Some decidable and undecidable statements, 2007

875 **C. Krattenthaler and T. Rivoal,** Hypergéométrie et fonction zêta de Riemann, 2007

874 **Sonia Natale,** Semisolvability of semisimple Hopf algebras of low dimension, 2007

873 **A. J. Duncan,** Exponential genus problems in one-relator products of groups, 2007

872 **Anthony V. Geramita, Tadahito Harima, Juan C. Migliore, and Yong Su Shin,** The Hilbert function of a level algebra, 2007

871 **Pascal Auscher,** On necessary and sufficient conditions for L^p-estimates of Riesz transforms associated to elliptic operators on $\mathbb{R}^n$ and related estimates, 2007

870 **Takuro Mochizuki,** Asymptotic behaviour of tame harmonic bundles and an application to pure twistor D-modules, Part 2, 2007

869 **Takuro Mochizuki,** Asymptotic behaviour of tame harmonic bundles and an application to pure twistor D-modules, Part 1, 2007

868 **Gelu Popescu,** Entropy and multivariable interpolation, 2006

867 **Vilmos Totik,** Metric properties of harmonic measures, 2006

866 **William Craig,** Semigroups underlying first-order logic, 2006

865 **Nathanial P. Brown,** Invariant means and finite representation theory of $C*$-algebras, 2006

864 **John M. Lee,** Fredholm operators and Einstein metrics on conformally compact manifolds, 2006

For a complete list of titles in this series, visit the
AMS Bookstore at **www.ams.org/bookstore/**.